BestMasters

Springer awards „BestMasters" to the best master's theses which have been completed at renowned Universities in Germany, Austria, and Switzerland.

The studies received highest marks and were recommended for publication by supervisors. They address current issues from various fields of research in natural sciences, psychology, technology, and economics.

The series addresses practitioners as well as scientists and, in particular, offers guidance for early stage researchers.

Carina Klein

Testing Modern Biostratigraphical Methods

Application to the Ammonoid Zonation across the Devonian-Carboniferous Boundary

 Springer Spektrum

Carina Klein
Berlin, Germany

BestMasters
ISBN 978-3-658-15344-1   ISBN 978-3-658-15345-8   (eBook)
DOI 10.1007/978-3-658-15345-8

Library of Congress Control Number: 2016948610

Springer Spektrum

Printed on acid-free paper

This Springer Spektrum imprint is published by Springer Nature
The registered company is Springer Fachmedien Wiesbaden GmbH
The registered company address is: Abraham-Lincoln-Strasse 46, 65189 Wiesbaden, Germany

Acknowledgements

First of all and most important, I want to thank my supervisors PD Dr. Dieter Korn and Prof. Dr. Michael Schudack for their advice. I would also like to thank Johan Renaudie for his advice concerning CONOP. Furthermore, I would like to thank Sonny A. Walton and Hanna Nowinski for proofreading.

Table of Contents

Index of Figures

Index of Tables

1 Abstract

The occurrences of 64 late Famennian (Late Devonian) ammonoid species from 13 sections and 52 early Tournaisian (Early Carboniferous) ammonoid species from 7 sections are ordered stratigraphically. Three stratigraphical correlation methods are used, (1) Unitary Associations (UA), (2) Constrained Optimization (CONOP) and (3) Ranking and Scaling (RASC) to test the quality of the existing modern ammonoid zonation and to see which of the three methods is best suitable for the refining of the currently used ammonoid zonation. The results obtained from these methods are compared with each other with respect to ammonoid succession and resolution; they were tested with the empirical data from selected reference sections.

Principally, the UA, CONOP and RASC methods lead to similar outcomes with respect to the succession of occurrence events of the analysed ammonoid species in the various sections. Additionally the fit with the reference sections is generally good. On the basis of the results of the three analyses, the existing modern ammonoid zonation can be confirmed for the Devonian dataset and partly refined for the Carboniferous dataset.

Which method is most suitable depends on the data available and the purpose of the investigation. For the biostratigraphical analysis, the RASC method is considered as the most suitable because the result perfectly mirrors the existing modern ammonoid zonation. The more conservative UA method facilitates the separation of zones. UA and RASC are recommended and can be used to complement one another. The CONOP approach is the least suitable, because the calculation takes a long time and the results do not mirror the existing modern ammonoid zonation.

2 Introduction

The application of modern biostratigraphical methods, especially concerning ammonoid stratigraphy, is a new and promising approach. These new methods provide a finer stratigraphical resolution and a minimization of contradictions without new extensive sampling efforts being required. The three biostratigraphical methods Unitary Associations (UA) (Guex 1991), Constrained Optimization (CONOP) (Kemple et al. 1989 and 1995) and Ranking and Scaling (RASC) (Gradstein and Agterberg 1982) are used for late Famennian (Late Devonian) and early Tournaisian (Early Carboniferous) ammonoid successions and evaluated with respect to their resolution and suitability. The methods were tested on published and unpublished data from various sources.

The Unitary Associations method was originally developed for the stratigraphical occurrences of ammonoids (Guex 1991). It leads to a noticeable improvement of the biochronological resolution for many groups, including the Ammonoidea. The resolution of ammonoid biostratigraphy has been be improved by means of the Unitary Associations method in several cases: Monnet and Bucher (2002, 2006) gained a higher resolution for Late Creataceous ammonoid biostratigraphy of Western Europe and for Middle Triassic North American ammonoid biostratigraphy respectively. Brühwiler et al. (2010) obtained an unprecedented high-resolution for the Early Triassic ammonoid biostratigraphy of India and Monnet et al. (2011) refined the late Emsian and Eifelian ammonoid biostratigraphy of Morocco. Cooper et al. (2001) compared the methods CONOP and RASC for Paleocene to early Miocene foraminifera, nannofossil, dinoflagellate and miospore occurences. They found that both techniques greatly improved biostratigraphical precision compared to conventional methods and showed that a higher quality result can be obtained from the same data.

Sections, which were sampled with a very fine resolution, are used to refine the latest Devonian and earliest Carboniferous ammonoid stratigraphy for the Rhenish Mountains, which is one of the classical regions for Devonian and Carboniferous ammonoid stratigraphy in particular the boundary between the two periods (Wedekind 1914; Schindewolf 1937; Vöhringer 1960; Korn 1993, 2002).

With 265 species occurring in Central Europe, the South Urals, North Africa etc. in the late Famennian and 100 species in the early Tournaisian (Korn and Klug 2012) ammonoids possess a high diversity. Therefore, the Ammonoidea are suitable candidates for the development of a stratigraphical scheme across the Devonian-Carboniferous boundary. In addition ammonoids are very abundant and morphologically diverse on both sides of the boundary and hence serve as excellent index fossils (Korn 1993). Three groups of ammonoids can be distinguished across the Devonian-Carboniferous boundary, the order Clymeniida with a dorsal sipuncle and the orders Goniatitida and Prolecanitida with a ventral siphuncle.

The goals of this MSc thesis are: (1) to refine the latest Devonian and earliest Carboniferous ammonoid stratigraphy for the Rhenish Mountains, (2) to investigate if the modern biostratigraphical methods UA, CONOP and RASC are suitable for this purpose and (3) if the methods are suitable, to test which of these methods is the best.

2.1   Historical background

A detailed historical review on the Late Devonian ammonoid biostratigraphy of the Rhenish Mountains was provided by Korn (2002). Von Buch (1832) was the first, who described the Late Devonian goniatite fauna near Adorf but a stratigraphical subdivision was not conducted. Kayser (1872, 1873) and later Denckmann (1894) proposed coarse stratigraphical schemes for the Late Devonian sedimentary succession. Denckmann (1901) proposed the lithostratigraphical units "Enkeberger Kalk", "Dasberger Kalk" and "Wocklumer Kalk", named after classical localities, of which the latter two are still used. Frech (1897, 1902) developed the biostratigraphical "Stufen" and "Zonen" scheme and Wedekind (1914, 1918) subdivided the Famennian (in the current definition) into five Stufen: *Cheiloceras*-Stufe, *Prolobites*-Stufe, the *Postprolobites*-Stufe, *Laevigata-Gonioclymenia*-Stufe and *Wocklumeria*-Stufe. This framework was refined by Lange (1929) using index fossils to demarcate the stratigraphical units more precisely.

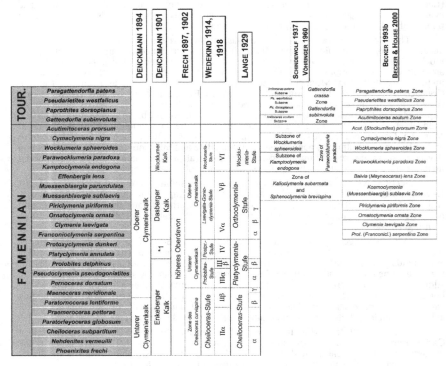

Fig. 1: Revised ammonoid zonation and historical subdivisions of the Late Devonian and Early Carboniferous rocks in the Rhenish Mountains [*1 = Zone der *Clymenia annulata*] (from Korn 2002).

Schindewolf (1937) investigated the Wocklum Limestone at the Oberrödinghausen railway cutting, which was, with a centimetre scale, the most detailed study of the Latest Devonian stratigraphy at that time. He recorded over 65 ammonoid species from the Wocklum Limestone, which he subdivided into three biostratigraphical units. Vöhringer (1960) investigated Early Carboniferous rocks at the same locality. He subdivided the Hangenberg Limestone into four biostratigraphical units with 45 species.

It was not until the 1980s that the ammonoids of this period became the focus of research once again. Korn (1981, 1986, 1992, 1994, 1995) and Becker (1996, 2000) conducted intense studies on the late Famennian ammonoid faunas. The correlation charts by Becker (1993) and Becker and House (1994, 2000) provided stratigraphical data with higher resolution (Fig. 1).

2.2 Regional geology

The Rhenish Mountains are part of the Rhenohercynian Zone of the Variscan orogenetic complex. Most of the localities of this MSc thesis are located on the northern flank of the Remscheid-Altena Anticline (Hasselbach, Becke-Oese, Oberrödinghausen) or on the anticlinal crest (Dasberg, Effenberg, Müssenberg). The Remscheid-Altena Anticline borders on the Lüdenscheid Syncline in the south, where the Stockum locality is located. The Drewer locality has a position on the Belecke Anticline, which plunges under the Late Cretaceous cover of south-eastern Münsterland. The Belecke Anticline can be subdivided into a western and an eastern part, of which the eastern part is completely disclosed in the Drewer locality (Clausen in Luppold et al. 1994).

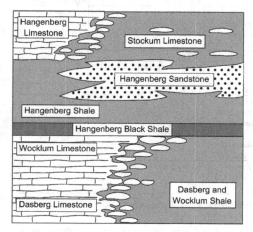

Fig. 2: Facies and lithology of the rise and the basin (from Clausen in Luppold et al. 1994).

The Remscheid-Altena Anticline is 10 to 15 km wide and plunges in a north-eastern direction. The late Famennian rock inventory is composed of red and green shales, black shales, shales with limestone nodules and nodular limestone beds with few sandstone intercalations. The sediments were deposited on a relief, which was produced by Middle and early Late Devonian reefs. Sections dominated by shales, siltstone and sandstone are characteristic for the basins; sections predominantly consisting of shales with limestone nodules and nodular flaser limestone are typical of slopes; cephalopod limestone sections are typical for submarine rises. Regressive trends caused sandstone intercalations in all of these palaeogeographical settings. Sections now located on the western part of the Remscheid-Altena Anticline (Becke-Oese and Hasselbach) had a bathymetric position deeper than those located on the eastern anticlinal crest. This is indicated by the increasing clay content towards the west (Clausen in Luppold et al. 1994; Fig. 2).

With decreasing clay content towards the east, the nodular limestone beds tend to amalgamate, as in the sections of Oberrödinghausen and Dasberg. The Müssenberg and Effenberg sections in the eastern Remscheid-Altena Anticline represent the shallowest part of the slope and are therefore those with the lowest shale contents (Clausen in Luppold et al. 1994; Korn and Weyer 2003).

The Drewer section resembles the Oberrödinghausen section and possibly had a similar bathymetric position. The nodular latest Devonian limestone beds contain only thin shaly interbeds. The locality of Stockum had a basinal palaeogeographical position with a rather thick siliciclastic succession (Luppold et al. 1994; Korn and Weyer 2003).

2.3   Lithological frame

Six lithological units contain the ammonoid succession, which is analysed in the following (Tab. 1; Fig. 3; Fig. 4).

Lithological unit	Stratigraphical unit	Thickness	Bed thickness
Hangenberg Limestone	*Paragattendorfia patens* Zone *Pseudarietites westfalicus* Zone *Paprothites dorsoplanus* Zone *Gattendorfia subinvoluta* Zone	1.5-2.5 m	5-10 cm
Stockum Limestone	*Acutimitoceras prorsum* Zone	20 cm	
Hangenberg Shale/ Sandstone	*Acutimitoceras prorsum* Zone	Hangenberg Shale: 4-6 m; Hangenberg Sandstone: 30 m	
Hangenberg Black Shale	*Cymaclymenia nigra* Zone	15-30 cm	
Wocklum Limestone	*Wocklumeria denckmanni* Zone *Parawocklumeria paradoxa* Zone *Kamptoclymenia endogona* Zone *Effenbergia lens* Zone *Muessenbiaergia parundulata* Zone *Muessenbiaergia sublaevis* Zone	3.5-6 m	5-20 cm
Dasberg Limestone	*Piriclymenia piriformis* Zone *Ornatoclymenia ornata* Zone *Clymenia laevigata* Zone	0.75-1.5 m	5-12 cm

Tab. 1: Lithological units in ascending order (with their biostratigraphical position).

The shelf carbonates across the Devonian-Carboniferous boundary are pelagic nodular limestones in the Rhenish Mountains (Luppold et al. 1994; Korn 2002). The fossiliferous limestones across the Devonian-Carboniferous boundary are:

(1) The Dasberg Limestone, which is grey to reddish in colour. The lower part is mainly composed of nodular limestone beds and shale horizons, towards the top, the limestone beds become more compact (Korn and Luppold 1987).

(2) The Wocklum Limestone, which is uniformly composed of alternating shales and nodular limestone beds, is dark grey to blue in colour. The carbonate concentration is irregularly distributed. A gradual decrease in shale content towards the top is observable. Within the Wocklum limestone 15 carbonate concentration peaks, which reflect limestone

beds or amalgamated nodular limestone beds, can be recognised.

They can be used for the correlation of the neighbouring sections (Korn and Weyer 2003). Up to three volcanic ash layers, so called metabentonite layers, with a maximum thickness of 2 cm are intercalated in the Wocklum Limestone at the localities Hasselbach and Becke-Oese. They can be used for detailed dating of the section as well as the correlation of neighbouring sections (Korn and Weyer 2003).

(3) The Stockum Limestone is only present at the Stockum locality, where a calcarenaceous accretion formed limestone beds and lenses in the Hangenberg Sandstone. This allochthonous limestone shows a gradation (Korn and Clausen in Clausen et al. 1994).

(4) The Hangenberg Limestone is composed of 35 nodular limestone beds with shaly interbeds. The clastic content decreases towards the top, the fossil content increases towards the top of the Hangenberg Limestone (Luppold et al. 1994). The Hangenberg Limestone features 13 carbonate peaks, which allow for a subdivision into 13 lithological units. The 13 lithological units are defined by cycles with a nodular limestone bed at the bottom, followed by limestone nodules in shales and pure shales at the top (Korn and Weyer 2003). The Hangenberg Limestone contains up to three metabentonite layers at the localities Hasselbach (max. 2 cm thick) and Becke-Oese (max. 20 cm thick). A metabentonite bed in the *Gattendorfia subinvoluta* Zone was dated to 353.2 ± 4.0 Ma by Claoué-Long et al. (1992) and later to 355.7 ± 4.2 Ma by Claoué-Long (1995). Trapp et al. (2004) determined an age of 360.5 ± 0.8 Ma for this horizon.

Fig. 3: Drewer locality (photo by Pitz 1971).

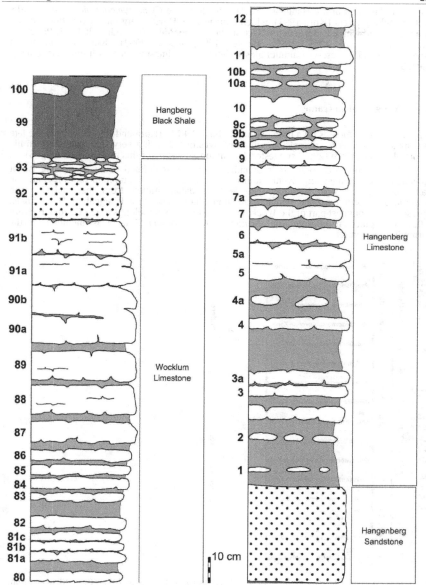

Fig. 4: Lithological log of the sections exemplified for the Drewer section (by Korn et al. 1994).

The pitch black and brittle Hangenberg Black Shale of the *Cymaclymenia nigra* Zone marks the Hangenberg Event (Korn 1991), which indicates the Hangenberg mass extinction with its faunal turnover, which had a severe effect on the Ammonoidea (Schindewolf 1937; Price and House 1984; Korn 1986, 1993; Becker 1993; House 2002). Korn (1991) found threedimensionally preserved ammonoids in bituminous limestone lenses in the Hangenberg Black Shale.

2.4 Stratigraphical frame

In total this study focuses on a time period of about 8.4 Ma (Korn and Ilg 2007) from the late Famennian to the early Tournaisian, which is investigated with a very fine resolution. In his historical review, Korn (2002) suggested 16 ammonoid zones for this time interval (Fig. 5).

The Devonian-Carboniferous boundary, which was defined using conodonts, is not equivalent to the Hangenberg Event but has a position about two ammonoid zones higher. The Hangenberg Event was, because of a lack of data, omitted from the analysis. The *Acutimitoceras prorsum* Zone, which occurs immediately above the Hangenberg Black Shale, represents the youngest part of the *Wocklumeria* Stufe and still belongs to the Devonian (Korn 2002). For practical reasons, the *Acutimitoceras prorsum* Zone was included in the Carboniferous dataset.

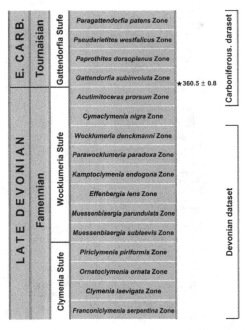

Fig. 5: Revised ammonoid zonation by Korn (2002) and dating by Trapp et al. (2004) (from Klein and Korn 2015).

2.5 Ammonoid diversity

The order Clymeniida experienced a strong diversification with more than 15 co-occurring genera at their maximum in the late Famennian ammonoid assemblages (Korn 1986; 1993). The Clymeniida experienced a major decline at the Hangenberg Event. Only two genera (*Cymaclymenia* and *Postclymenia*) survived (Korn 1993), which became extinct a short time after the event (Korn et al. 2004). The order Goniatitida, which originated already in the Middle Devonian, also experienced a major decline (Becker 1993). Although the Goniatitida are much less abundant in the Famennian than the Clymeniida (Korn 1993), they survived the extinction event and flourished in the Carboniferous and Permian.

The species richness in the Rhenish Mountains increases until the *Kamptoclymenia endogona* Zone and afterwards decreases again towards the Hangenberg Event (Tab 2). After the Hangenberg Event, in the earliest Carboniferous, the ammonoids radiated rapidly (Vöhringer 1960; Korn 1993). The ammonoids radiating in the early Carboniferous are descendants of *Acutimitoceras* and the appearance of *Eocanites* marks the beginning of the order Prolecanitida (Vöhringer 1960). The diversity pattern in the Rhenish Mountains does not change much throughout the Earliest Carboniferous. A rapid increase of species richness is noticeable already a short time after the Hagenberg Event. Only at the end of the *Gattendorfia* Stufe, a slight decline in species richness can be seen (Tab. 3).

Ammonoid zone	Species richness
Clymenia laevigata	7
Ornatoclymenia ornata	8
Piriclymenia piriformis	7
Muessenbiaergia sublaevis	16
Muessenbiaergia parundulata	17
Effenbergia lens	26
Kamptoclymenia endogona	29
Parawocklumeria paradoxa	25
Wocklumeria denckmanni	16

Tab 2: Devonian species richness.

Ammonoid zone	Species richness
Acutimitoceras prorsum	9
Gattendorfia subinvoluta	21
Paprothites dorsoplanus	20
Weyerella molaris	23
Pseudarietites westfalicus	18
Paragattendorfia patens	23

Tab 3: Carboniferous species richness.

3 Material

The data used in this study consists of data from literature sources as well as unpublished data. The unpublished data comes from collections stored at the Museum für Naturkunde Berlin. The ammonoid species from the unpublished collections were determined by Dieter Korn, who also provided the columnar sections.

3.1 Fossil species

The Devonian and Carboniferous datasets contain a total of 64 and 52 ammonoid species, respectively. The original datasets needed to be corrected to improve the results of the Unitary Associations method: Species with open nomenclature were omitted to avoid stratigraphical uncertainties caused by taxonomic uncertainties (Monnet et al. 2011). The datasets were also checked for taxa, whose occurrence can not be explained in some horizons and were therefore either incorrectly identified or incorrectly assigned to the horizon. Finally, taxa, which exclusively define a UA, which does not coincide with the empirical stratigraphical ordering, were deleted, because they do not provide information for correlation and also add uncertainties to the datasets.

Devonian

The Devonian dataset (which includes the *Clymenia laevigata* Zone to the *Wocklumeria denckmanni* Zone) contains 27 genera (Tab. 4).

Order	Suborder	Superfamily	Family	Genus	Species	n
Goniatitida	Tornoceratina	Priono-ceratoidea	Posttorno-ceratidae	*Discoclymenia*	*cucullata*	4
			Prionoceratidae	*Mimimitoceras*	*alternum*	2
					fuerstenbergi	8
					geminum	32
					lentum	19
					lineare	6
					liratum	50
					nageli	6
					pompeckji	1
					rotersi	2
					trizonatum	10
				Effenbergia	*falx*	39
					lens	32
					minutula	13
				Kenseyoceras	*biforme*	14
					nucleus	25
				Balvia	*globulare*	11
						...

Order	Suborder	Superfamily	Family	Genus	Species	n
...						
Clymeniida	Clymeniina	Platy-clymeniaceae	Platy-clymeniidae	Progonio-clymenia	acuticostata	1
			Piriclymeniidae	Piriclymenia	piriformis	6
				Ornatoclymenia	ornata	3
			Glatziellidae	Glatziella	glaucopis	19
				Soliclymenia	paradoxa	2
				Postglatziella	carinata	11
		Clymeniaceae	Clymeniidae	Clymenia	laevigata	10
			Kosmo-clymeniidae	Kosmoclymenia	effenbergensis	8
					inaequistriata	15
					lamellosa	1
					schindewolfi	30
					undulata	19
				Lissoclymenia	wocklumeri	37
				Muessenbergia	ademmeri	5
					bisulcata	8
					coronata	3
					diversa	1
					galeata	7
					parundulata	9
					sublaevis	58
					xenostriata	3
				Linguaclymenia	clauseni	33
					similis	98
		Wocklumeria-ceae	Para-wocklumeriidae	Kamptoclymenia	endogona	9
					trigona	3
				Parawocklumeria	distorta	11
					paprothae	23
					paradoxa	40
					patens	11
			Wocklumeriidae	Wocklumeria	denckmanni	34
		Gonio-clymeniaceae	Gonio-clymeniidae	Kalloclymenia	pessoides	3
					subarmata	32
					uhligi	3
				Finiclymenia	wocklumensis	41
				Gonioclymenia	speciosa	10
			Sellaclymeniidae	Sellaclymenia	torleyi	1
	Cyrto-clymeniina	Cyrto-clymeniaceae	Cyrtoclymeniidae	Cyrtoclymenia	angustiseptata	44
					plicata	4
			Cymaclymeniidae	Cymaclymenia	camerata	2
					cordata	13
					costellata	32
					curvicosta	1
					involvens	10
					striata	109
					tricarinata	1
						...

Order	Suborder	Superfamily	Family	Genus	Species	n
...						
					warsteinensis	20
				Rodachia	dorsocostata	2

Tab. 4: Taxa of the Devonian dataset [n=number of occurrences].

Carboniferous

The Carboniferous dataset (which includes the latest Devonian *Acutimitoceras prorsum* Zone to the Carboniferous *Paragattendorfia patens* Zone) contains 17 genera (Tab. 5).

Order	Suborder	Superfamily	Family	Genus	Species	n
Goniatitida	Tornoceratina	Priono-cerataceae	Prionoceratidae	Mimimitoceras	hoennense	7
					varicosum	4
				Globimitoceras	globiforme	9
				Paragattendorfia	globiformis	7
					patens	1
				Acutimitoceras	acutum	9
					antecedens	7
					convexum	3
					depressum	4
					exile	4
					intermedium	15
					kleinerae	11
					prorsum	2
					procedens	1
					simile	5
					stockumense	2
					subbilobatum	14
					undulatum	3
				Costimitoceras	ornatum	3
				Hasselbachia	gracilis	2
					multisulcata	7
					sphaeroidalis	10
				Nicimitoceras	acre	3
					caesari	1
					carinatum	4
					heterolobatum	9
					subacre	7
					trochiforme	9
				Voehringerites	peracutus	4
				Paralytoceras	serratum	1
				Paprothites	dorsoplanus	13
					raricostatus	1
					ruzhencevi	1
				Pseudoarietites	planissimus	1
					subtilis	3
					westfalicus	9
			Gattendorfiidae	Gattendorfia	costata	14
					crassa	5
					subinvoluta	9
						...

Order	Suborder	Superfamily	Family	Genus	Species	n
..					tenuis	11
				Kazakhstania	evoluta	2
				Weyerella	concava	5
					molaris	13
					reticulum	3
Prolecanitida	Prolecanitina	Prole-canitaceae	Prolecanitidae	Eocanites	brevis	4
					carinatus	1
					nodosus	13
					planus	1
					spiratissimus	1
					supradevonicus	4
					tener	2
Clymeniida	Clymeniina	Cyrto-clymeniaceae	Cymaclymeniidae	Postclymenia	evoluta	3

Tab. 5: Taxa of the Carboniferous dataset [n=number of occurrences].

3.2 Localities

The Late Devonian dataset includes 13 sections (Tab. 6); the Early Carboniferous dataset includes 7 sections (Tab. 7) in a total of 8 localities (Fig. 6).

Section	Short	Reference	n species	n horizons
Oberrödinghausen railway cutting	ORBK	Korn (unpublished)	21 species	13 horizons
Oberrödinghausen road cutting	ORSK	Korn (unpublished)	34 species	13 horizons
Oberrödinghausen road cutting alpha	ORSTA	Thiem (unpublished)	29 species	29 horizons
Oberrödinghausen road cutting beta	ORSTB	Thiem (unpublished)	30 species	27 horizons
Effenberg 1977	E77	Korn (unpublished)	27 species	15 horizons
Effenberg 1987	E87	Korn (unpublished)	7 species	16 horizons
Müssenberg 1	M1	Korn (unpublished)	44 species	78 horizons
Müssenberg 3	M3	Korn (unpublished)	9 species	7 horizons
Müssenberg 4	M4	Korn (unpublished)	23 species	20 horizons
Dasberg Middle	DASM	Korn (unpublished)	13 species	9 horizons
Dasberg North	DASN	Korn (unpublished)	5 species	10 horizons
Dasberg South	DASS	Korn (unpublished)	30 species	41 horizons
Drewer	DD	Korn et al. (1994)	11 species	7 horizons

Tab. 6: The Late Devonian sections with number of species and number of horizons.

Section	Short	Reference	n species	n horizons
Hasselbach	H	Korn and Weyer (2003)	20 species	16 horizons
Becke-Oese	BO	Korn and Weyer (2003)	9 species	6 horizons
Oberrödinghausen railway cutting	ORBW	Weyer (unpublished)	23 species	15 horizons
Oberrödinghausen railway cutting	ORBV	Vöhringer (1960)	45 species	10 horizons
Müssenberg	M2	Korn (unpublished)	7 species	3 horizons
				...

Section	Short	Reference	n species	n horizons
...				
Stockum	SK	Korn (1984)	7 species	1 horizon
Drewer	DK	Korn et al. (1994)	7 species	8 horizons

Tab. 7: The Early Carboniferous sections with number of species and number of horizons.

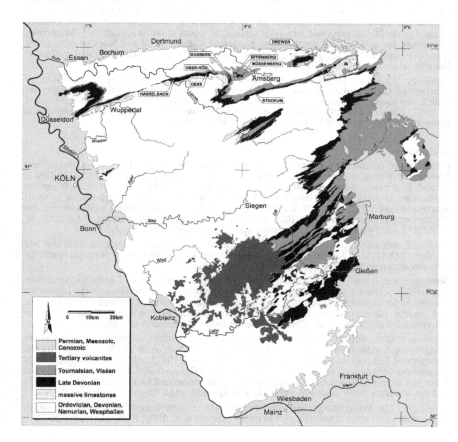

Fig. 6: Geographical positions of section localities (from Korn 2002).

Hasselbach

Geographical position: The Hasselbach section is located at the banks of the Hassel rivulet between Hagen-Reh and Hagen-Henkhausen (51.37388° N, 7.57247° E).

Analysed section: Only the Carboniferous part of the section is analysed with the bed numbers H-83 to H-45 in ascending order. Twenty species compose the total data set, of which 16 are represented by a single occurrence. Two species (*Acutimitoceras intermedium* and *Paprothites dorsoplanus*) are best represented and occur in four horizons.

Stratigraphical frame: *Gattendorfia subinvoluta* Zone to *Pseudarietites westfalicus* Zone.

Previous studies: Schmidt (1924) was the first, who described the rock succession of the outcrop. He distinguished between the Dasberg and Wocklum limestones, the Hangenberg Black Shale and the Hangenberg Limestone. Becker et al. (1984) used conodonts and ammonoids for a detailed description of the Hangenberg Limestone and the Hangenberg Shale. Becker (1988) presented columnar sections of a part of the Wocklum and Hangenberg limestones. He misleadingly postulated an overlap of *Kalloclymenia* and *Wocklumeria*, which was later corrected by Luppold et al. (1994). Korn and Weyer (2003) provided a detailed lithological description of the section.

Lithology: Only the Hangenberg Limestone is used in this study. It is composed of interbedded thin nodular limestones, which contain only a limited number of fossils, and shales. Metabentonite horizons can be used for detailed dating of the section. (Claoué-Long et al. 1992; Claoué-Long 1995; Trapp et al. 2004). The base of the Hangenberg Limestone is formed by a platy turbiditic limestone (Korn and Weyer 2003).

Becke-Oese

Geographical position: Three outcrops at the western side of the B7 between Hemer and Menden have been combined across the Devonian-Carboniferous boundary beds; the first is located in a road cutting (51.40082° N, 7.78680° E) and displays the Devonian limestone formations as well as the Hangenberg Black Shale, the second is exposed in a small abandoned quarry (51.40106° N, 7.78704° E) and exposes the Hangenberg Sandstone and the Hangenberg Limestone and the third is again located at the road cutting (51.40132° N, 7.78751° E) and exposes the Hangenberg Limestone and following formations.

Analysed sections: One combined section of the three outcrops is analysed with the bed numbers BO-11 to BO-37 in ascending order. Eight of the nine species are represented by only one occurrence and the ninth species by 2 occurrences.

Stratigraphical frame: The compound section exposes a complete late Famennian to middle Tournaisian section. Only the Early Carboniferous ammonoids of the Hangenberg Limestone (*Gattendorfia subinvoluta* Zone to *Paragattendorfia patens* Zone) were used in this analysis. The sparse occurrence of ammonoids and conodonts does not allow for a precise determination of the Devonian-Carboniferous boundary (Luppold et al. 1994).

Previous studies: Schmidt (1924) presented the first lithological subdivision: He found the Hangenberg Sandstone as a thick bedded greywacke with shale at its top, followed by the Hangenberg Limestone. Kullmann (in Paproth and Streel 1982) published a compilation of the ammonoid faunas of the Hangenberg Limestone in this section. Luppold et al. (1994) and Korn and Weyer (2003) published descriptions of the ammonoid fauna of the *Clymenia* Stufe and the *Wocklum* Stufe as well as a conodont stratigraphy and a facies analysis.

Lithology: The Hangenberg Limestone begins with a platy limestone bed, which shows characteristics of a distal turbidite rich in mica (Luppold et al. 1994; Korn and Weyer 2003).

Oberrödinghausen railway cutting

Geographical position: The Oberrödinghausen railway cutting section is located in the Hönne Valley between Menden and Balve, at the western margin of the large cement works of Ober-Rödinghausen (51.39429° N, 7.84113° E).

Analysed sections:

(1) The unpublished bed-by-bed sampling of the section by Korn (ORBK; bed numbers ORBK-11B to ORBK-1A in ascending order) from the eastern side of the railway cutting.

(2) The unpublished bed-by-bed sampling of the section by Weyer (ORBW; bed numbers ORBW-6b to ORBW-2a in ascending order) from the eastern side of the railway cutting.

(3) The bed-by-bed sampling of the section by Vöhringer (1960) (ORBV; bed numbers ORBV-6 to ORBV-1 in ascending order) from the western side of the railway cutting.

ORBV is the best of the Carboniferous sections in this analysis and is hence used as a reference. The section sampled by Weyer was taken only about five meters away from Vöhringer's section. Vöhringer (1960) subdivided the section into six beds, of which he subdivided bed 3 into five subunits. Weyer used the same bed numbering. He did not sample bed 1, but subdivided beds 2, 3c, 3d, 4 and 6 into two subunits and bed 5 into four subunits. Weyer's section has a higher resolution, but his collections were smaller and hence less diverse than Vöhringer's collections.

Stratigraphical frame:

ORBK: Wocklum Limestone from the *Kamptoclymenia endogona* Zone to the *Wocklumeria denckmani* Zone

ORBW and ORBV: Hangenberg Limestone from the *Gattendorfia subinvoluta* Zone to the *Paragattendorfia patens* Zone (Fig. 7)

Previous studies: This famous locality is, for fossils, one of the richest Devonian-Carboniferous boundary sections worldwide. Schmidt (1924) found a rich cephalopod assemblage in the Hangenberg Limestone. The global subdivisions of the *Wocklumeria* and *Gattendorfia* Stufen are based on investigations of this section (Schindewolf 1937; Vöhringer 1960). Luppold et al. (1994) found ammonoids in the Hangenberg Black Shale.

Lithology: The Wocklum Limestone and the Hangenberg Limestone are composed of alternating shales and nodular limestone (Korn and Weyer 2003). The clay content of the Hangenberg Limestone is remarkably high.

Oberrödinghausen road cutting

Geographical position: The Oberrödinghausen road cutting section is located at the eastern side of the Hönne Valley between Menden and Balve, east of the large cement works of Ober-Rödinghausen (51.39385° N, 7.84478° E).

(1) The unpublished bed-by-bed sampling of the section by Korn (ORSK; bed numbers ORSK-20 to ORSK-1 in ascending order)

(2) The unpublished bed-by-bed sampling of the section by Thiem (ORSTA; bed numbers ORSTA-7a(3) to ORSTA-1(1) in ascending order)

(3) The unpublished bed-by-bed sampling of the section by Thiem (ORSTB; bed numbers ORSTB-12(4) to ORSTB-6b(1) in ascending order)

ORSTB represents the lower part of the succession, ORSTA the upper part, with an overlapping part in the middle.

Stratigraphical frame:

ORSK: *Kamptoclymenia endogona* Zone to *Wocklumeria denckmani* Zone

ORSTA: *Effenbergia lens* Zone to *Wocklumeria denckmani* Zone (Fig. 8)

ORSTB: *Effenbergia lens* Zone to *Kamptoclymenia endogona* Zone

Lithology: The difference between the Oberrödinghausen railway cutting and the Oberrödinghausen road cutting is the presence of the approximately 15 m thick Hagenberg Shale, with intercalated thick-bedded sandstone beds, which belong to the Hangenberg Sandstone (Luppold et al. 1994; Korn and Weyer 2003).

Previous studies: Ziegler (1962) was the first to publish a detailed description of the Late

Devonian sections and its conodont fauna. Thiem and later Korn sampled this locality intensively for ammonoids, but the results were never published. Luppold et al. (1994) found the Wocklum Limestone to be rich in ammonoids and the Hangenberg Limestone moderately rich in ammonoids. Korn and Weyer (2003) noticed a close resemblance to the Oberrödinghausen railway cutting section, which is located approximately 200 meters away.

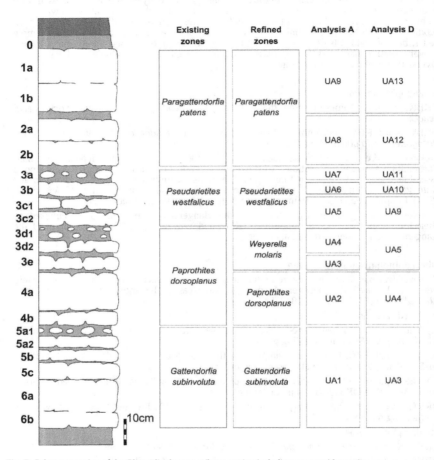

Fig. 7: Columnar section of the Oberrödinghausen railway cutting including ammonoid zonation.

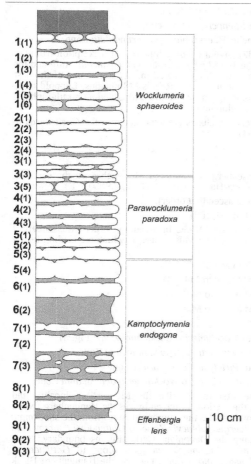

Fig. 8: Columnar section of the Oberrödinghausen road cutting including ammonoid zonation.

Effenberg

Geographical position: The Effenberg section is located at the large active quarry on the Effenberg northwest of Hachen (51.39080° N, 7.96120° E).

Analysed sections:

(1) The unpublished bed-by-bed sampling of the section by Korn in 1977 (E77; bed numbers E77-B to E77-V in ascending order)

(2) The unpublished bed-by-bed sampling of the section by Korn in 1987 (E87; bed numbers E87-J4 to E87-R in ascending order)

Stratigraphical frame: *Clymenia* Stufe and *Wocklumeria* Stufe

Effenberg 77: *Clymenia laevigata* Zone to *Effenbergia lens* Zone

Effenberg 87: *Piriclymenia piriformis* Zone to *Muessenbiaergia parundulata* Zone

Previous studies: Korn and Luppold (1987) gave a detailed lithological description from the *Cheiloceras* Stufe to the middle part of the *Wocklumeria* Stufe. The *Clymenia* Stufe and the *Wocklumeria* Stufe are rich in fossils. The authors considered the section as one of the best sections for Late Devonian cephalopod limestones in the Rhenish Mountains. Luppold et al. (1994) described the section in detail, including an ammonoid and conodont stratigraphy, a microfacies analysis and the bathymetric development.

Lithology: Nodular limestones with intercalated shales are exposed from the *Cheiloceras* to the *Gattendorfia* Stufe (Luppold et al. 1994).

Müssenberg

Geographical position: The locality Müssenberg is composed of trenches on the southern slope of the Müssenberg, 1.2 km N of Sundern-Hachen (51.38831° N, 7.98502° E).

Analysed sections (bed numbers are given in ascending order):

(1) Müssenberg 1 (M1; bed numbers M1-4 to M1-109)

(2) Müssenberg 2 (M2; bed numbers M2-3A to M2-3C in ascending order; Müssenberg 2 represents the three youngest horizons of the section Müssenberg 1, which are Carboniferous in age)

(3) Müssenberg 3 (M3; bed numbers M3-7 to M3-1)

(4) Müssenberg 4A (M4A; bed numbers M4A-18 to M4A-4)

(5) Müssenberg 4B (M4B; bed numbers M4B-16 to M4B-1)

(6) Müssenberg 4C (M4C; bed numbers M4C-15 to M4C-8)

Stratigraphical frame:

Müssenberg 1: *Clymenia laevigata* Zone to *Wocklumeria denckmanni* Zone (Fig. 9)

Müssenberg 2: *Acutimitoceras prorsum* Zone to *Gattendorfia subinvoluta* Zone

Müssenberg 3: *Clymenia laevigata* Zone to *Piriclymenia piriformis* Zone

Müssenberg 4A-4C: *Kamptoclymenia endogona* Zone to *Wocklumeria denckmanni* Zone

Previous studies: Korn (1981) was the first to describe the Müssenberg locality and particularly investigated the small part of the section, which embraces the Hangenberg Event. Luppold et al. (1984) stated that the section includes the ammonoid zones from the *subarmata* Zone to the *subinvoluta* Zone (*Muessenbiaergia sublaevis* Zone to *Gattendorfia subinvoluta* Zone in new terminology), hence bridging the Devonian- Carboniferous boundary. Korn (2002) found all ammonoid zones from the *Clymenia laevigata* Zone to *Gattendorfia subinvoluta* Zone except for the *Cymaclymenia nigra* Zone, because the Hangenberg Black Shale is absent.

Lithology: The limestone succession appears to be complete without major gaps. Only a few shale and siltstone layers interrupt the carbonate sedimentation. The lithofacies and carbonate microfacies were investigated by Luppold et al. (1994); they showed that biomicrites and biosparites dominate the lower part of the section, while biomicrudites and biosparrudites are more common in the upper part. The presence of micrites indicates calm and deep water conditions during deposition.

Dasberg

Geographical position: The locality Dasberg contains trenches and a road cutting 1 km WNW of Hövel (Dasberg S: 51.37174° N, 7.91031° E; Dasberg M: 51.37211° N, 7.91029° E;

Dasberg N: 51.37229° N, 7.91010° E).

Analysed sections: Three unpublished bed-by-bed samplings of the section by Korn are analysed:

(1) Dasberg Middle (DASM; bed numbers DASM-6 to DASM-1 in ascending order; represents the upper part of the succession)

(2) Dasberg North (DASN; bed numbers DASN-9 to DASN-26 in ascending order; represents the lower part of the succession)

(3) Dasberg South (DASS; bed numbers DASS-55 to DASS-1 in ascending order; represents the medium part of the succession)

Stratigraphical frame:

Dasberg Middle: *Parawocklumeria paradoxa* Zone to *Wocklumeria denckmani* Zone

Dasberg North: *Clymeina laevigata* Zone to *Ornatoclymenia ornata* Zone

Dasberg South: *Ornatoclymenia ornata* Zone to *Kamptoclymenia endogona* Zone (Fig. 10)

Previous studies: The mapping geologist Denckmann (1901) was the first author, who noted the fossil-rich nodular limestones in the Dasberg area. Wedekind (1914) distinguished ammonoid species at this locality. Because of the disadvantageous outcrop conditions, some authors (Schmidt 1924; Lange 1929) only collected surface material. In the late 1980s Korn dug trenches, where he collected 1600 late Famennian ammonoid specimens. Korn and Luppold (1987) subdivided the section into a lower part (*Clymenia* Stufe), which is rich in shale, and an upper part *(Muessenbiaergia sublaevis* Zone to *Effenbergia lens* Zone), which is poor in shale. Korn (2002) found all ammonoid zones from the *Francoclymenia serpentina* Zone to the *Kamptoclymenia endogona* Zone.

Lithology: The lithological development is similar to the neighbouring sections Effenberg and Müssenberg.

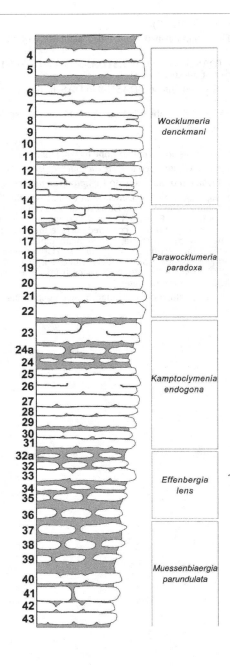

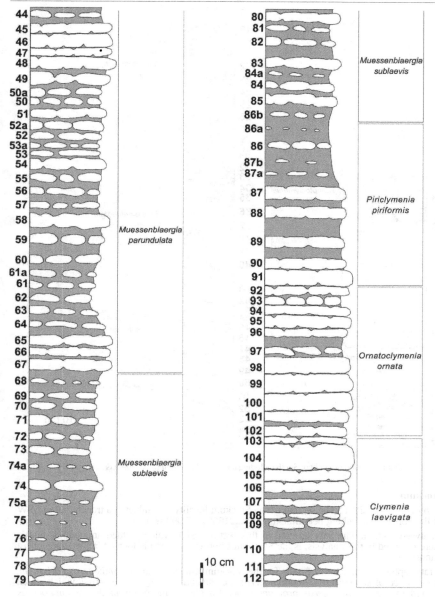

Fig. 9: Columnar section of Müssenberg including ammonoid zonation (by Luppuld et al. 1994).

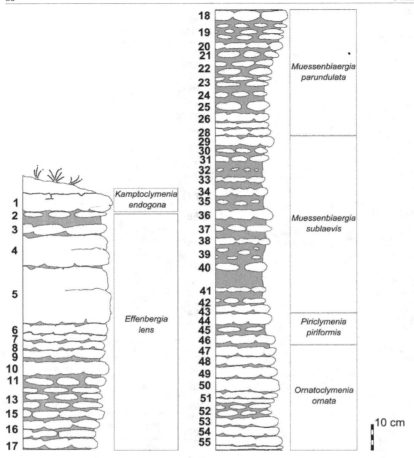

Fig. 10: Columnar section of Dasberg South including ammonoid zonation (by Clausen & Korn 2008).

Stockum

Geographical position: The section of the Stockum locality was taken at a trench 250 m SSW of the Spitzer Kahlenberg near Stockum (51.29185° N, 7.99832°).

Analysed section: The only horizon in this section (SK), which yielded ammonoids and is therefore used in this analysis, is the Stockum Limestone, which is located in the Hangenberg Sandstone.

Stratigraphical frame: Index fossils for the Stockum Limestone are Nici*mitoceras carinatum*, *Acutimitoceras stockumense*, *Acutimitoceras prosum* and *Nicimitoceras caesari*, which are all limited to the *Acutimitoceras prorsum* Zone. No clymeniids occur in the *Acutimitoceras* *prorsum* Zone.

Previous studies: The calcareous lenses, found at Stockum, containing goniatites were first described by Henke (1924) and Schmidt (1924). Vöhringer (1960) pointed out that the lenses are older than the Hangenberg Limestone. The term Stockum Limestone was coined by Alberti et al. (1974). Korn (1984) found numerous well preserved species of the genus *Acutimitoceras* at this locality. Clausen et al. (1994) provided a detailed biostratigraphy based on ammonoids, trilobites, ostracods, conodonts and spores.

Lithology: The complete section is mainly composed of shales, siltstone and sandstone with intercalated limestone nodules (Clausen et al. 1994). The limestone lenses of Stockum are grey in colour up to 8 cm thick and 50 cm in diameter.

Drewer

Geographical position: The section Drewer was examined at the north-western face of the abandoned Provinzialsteinbruch Drewer 40 m south the road between Belecke and Drewer (51.49430° N, 8.35680° E).

Analysed sections:

(1) Devonian part of the section (DD; bed numbers DD-1,2 to DD-93 in ascending order)

(2) Carboniferous part of the section (DK; bed numbers from DK-99 to DK-5 in ascending order)

Stratigraphical frame:

Devonian part of the section: *Wocklumeria* Stufe

Carboniferous part of the section: *Gattendorfia* Stufe

Previous studies: Schmidt (1922) was the first, who described this section and Ziegler (1971) presented a lithological section. Korn et al. (1994) found *Postclymenia evoluta* above the Hangenberg Black Shale in the *Acutimitoceras prorsum* Zone.

Lithology: The north-western face of the quarry shows silty mudstones with calcareous nodules and thinly bedded nodular limestones. Fine biomicrites are proof of a calm sedimentation environment, where the fine components were not washed away by currents. The nodular Dasberg and Wocklum limestones are bluish grey in colour. The Hangenberg Black Shale is overlain by the Hangenberg Sandstone (Korn et al. 1994).

4 Methods

I used methods of quantitative stratigraphy to rank, order and scale fossil events and ranges. The manual determination of the fossil succession can be difficult, especially when many sections with contradicting stratigraphic relationships (i.e. section 1: taxon A under taxon B; section 2: taxon B under taxon A) are involved. The manual determination of the fossil succession is time-consuming and leads to unreproducible results as each researcher can, for example, choose different index fossils. The automatised methods of quantitative stratigraphy are more reproducible. Two researchers will come to the same result using the same dataset and the same method (and the same adjustments). Still, it is possible to set subjective preferences by the use of different adjustments to the methods as well as to the datasets if desired. Depending on the method and the technical equipment, quantitative stratigraphic methods can be carried out much quicker than the traditional manual biostratigraphical work (Hammer and Harper 2006).

Three biostratigraphical approaches are applied to a Late Devonian and an Early Carboniferous dataset (Tab. 8):

(1) Unitary Associations (UA)

(2) Constrained Optimization (CONOP)

(3) Ranking and Scaling (RASC)

Reference sections are chosen in order to gain comparable results.

UA	CONOP	RASC
Program PAST	Program CONOP9	Program PAST
Deterministic	Deterministic	Probabilistic
Uses presence / absence matrix	Uses event order, best FADs and LADs	Uses event order, FADs and / or LADs are possible
Processes large datasets fast	Processes large datasets slowly	Processes large datasets fast
Treats all sections and events simultaneously	Treats all sections and events simultaneously	Treats all sections and events simultaneously
Produces a relative biochronological scale, which gives a sequence of intervals with unknown duration	Attempts to find maximum or most common stratigraphical ranges of taxa	Finds average stratigraphical postion of FADs and / or LADs
Uses the maximal coexistences of taxa to build associations	Uses simulated annealing to find the best composite sequence of events	Uses scores of event order relationships to find their most likely order
Discrete relative spacing of associations, where all original co-occurrences are preserved	Relative spacing of events in the composite is derived from the original event spacing	Relative spacing of events in the scaled optimum sequence
Solving of conflicting stratigraphical relationships by the "majority rule" and cycles by the "weakest link rule"	Numerous stratigraphical tests and graphical analyses of the stratigraphical result	Calculates standard deviation of each event as a function of its stratigraphical scatter

Tab. 8: Differences and similarities of the Unitary Associations, Constrained Optimization and Ranking and Scaling methods (Abbreviated and adapted from Gradstein in Gradstein et al. 2012).

It is important to note that every biostratigraphical method and also the reference datasets lead to their own first appearance dates (FADs) and last appearance dates (LADs). So the expressions FAD and LAD are not used as equivalent for the different results. Different FADs and LADs occur for the UA method, the CONOP method and the RASC method because of different resolutions of the results.

4.1   Unitary Associations (UA)

Guex (1991) introduced the method of Unitary Associations. This purely deterministic approach is well-suitable to establish biostratigraphical zonations in large and/or contradictory datasets (Monnet et al. 2011). The result is a relative biochronological scale, which gives a sequence of intervals with unknown duration (Guex 1991; Gradstein in Gradstein et al. 2012). The major difference between the UA approach and the CONOP and RASC approaches is that the UA method is based on associations, whereas the others are based on events. A range-through assumption is made and all co-occurences of taxa, either virtual or observed, are taken into account. The UA method is considered as conservative; robustness of the result is considered to be more important than the stratigraphical resolution. Uncertainties and contradictions are strongly considered, but confidence intervals are not calculated, the result is a stratigraphical succession of Unitary Associations (UAs). A UA is the maximal set of taxa, which can not be included in a larger set of taxa. It is based on the maximal co-occurences of these taxa (Hammer and Harper 2006; Gradstein in Gradstein et al. 2012).

The implementation of UA in the program PAST version 2.15 was used (Hammer and Harper 2001), which includes most of the features of the original program BioGraph (Savary and Guex 1999) and also some improvements (Hammer 2012). A presence/absence matrix is necessary with the species in columns in the different horizons of the stratigraphical sections in rows (Hammer and Harper 2006; Hammer 2012). For description of the steps of the method see Guex (1991), Monnet et al. (2011) and Hammer (2012).

The calculation of the UAs was carried out with different datasets to test which size of dataset and also which modifications lead to the best result. In Analysis A to Analysis C more sections were added to the reference section to test whether more collecting effort leads to a higher resolution and a better fit to the existing modern ammonoid zonation or to correlation problems and therefore a worse fit with the existing modern ammonoid zonation. In Analysis D, the complete datasets were analysed. The analysis of the complete dataset was also carried out with first occurrences only (Analysis E), which corresponds to traditional stratigraphical approaches ("using only the FADs checkbox" in PAST). A possibility to increase the robustness of the analysis is to omit singletons ("Null endemic tax" checkbox in PAST) (Analysis F). Those are per definition taxa, which only occur in one section (Monnet et al. 2011) , they increase the amount of contradictions, and are not helpful for the correlation of the sections (Monnet and Bucher 1999). Finally, the analysis was carried out at the genus level (Analysis G).

4.2   Constrained Optimization (CONOP)

Kemple et al. (1989 and 1995) developed the Constrained Optimization method. It is used to correlate stratigraphical events in a number of sections (Hammer and Harper 2006). The term Constrained Optimization refers to: Constrained, because it eliminates impossible solutions, and optimization, because it then searches for the best solution. CONOP only takes guesses about the best solution, which are improved iteratively, by comparing it with the original data. With a method called "inversion", the difference between the guess and the original data leads to the next guess (Gradstein in Gradstein et al. 2012). Because the time for the next guess increases by 2^N (N=number of events), CONOP uses "simulated annealing" to reduce search time (Sadler 2009; Gradstein in Gradstein et al. 2012). Like UA, CONOP is a deterministic approach, i.e., it computes the maximum stratigraphical range of a taxon, tops and bottoms are extended. Deterministic approaches assume that gaps occur due to missing data and that there is a true order of events. In contrast, probabilistic approaches assume that gaps occur due to random deviations (Gradstein in Gradstein et al. 2012).

The implementation of CONOP in PAST yielded results, which did not agree with the existing modern ammonoid zonation. Therefore, I used the program CONOP9, which can be downloaded at the iLearn of the UC Riverside (Regents of the University of California 2011). The data input for CONOP contains three files, the .SCT file, which labels the sections, the .EVT file, which labels the taxa and the .DAT file, which contains information about the FAD and LAD of all taxa in all sections. The three files were built according to the CONOP9 Quickstart Guide for Excel Users, which can also be downloaded at the iLearn of the UC Riverside (Regents of the University of California 2011). Unfortunately the manual data input takes a long time and it is prone to mistakes. Therefore, I used the R package CONOP9companion by Renaudie (2013) to help prepare my input files. The .CFG file was prepared after the Beginner's Command Syntax for CONOP9.CFG, which can also be downloaded at the iLearn of the UC Riverside (Regents of the University of California 2011). CONOP9 provides several output files, the most important being the compst.dat, which stores the composite section.

4.3 Ranking and Scaling (RASC)

Gradstein and Agterberg (1982) developed the method of Ranking and Scaling. The purpose of this method is a biostratigraphical ranking of events observed in different sections and also to give an estimate about the variability of the position of the events (Agterberg and Gradstein 1999; Hammer and Harper 2006). Inconsistencies can be solved by statistical averaging and stratigraphical reasoning (Agterberg and Gradstein 1999). The resulting ranked optimum sequence, including error bars, shows the biozonation and its precision. Unfortunately not all co-occurences are preserved (Hammer and Harper 2006). In contrast to UA and CONOP, RASC is a probalistic approach, i.e. it seeks to find the most probable or average range instead of the maximum range (Hammer and Harper 2006; Gradstein in Gradstein et al. 2012). If a species mostly shows short ranges, but a long range in one locality, still the shorter range will be assumed, this makes the method precise, but also may produce errors. Additionally there is no information given about global first and last occurrences (Hammer and Harper 2006). On the other hand, in this probabilistic approach, the most probable order of events in one location can be expected in other locations, too (Gradstein in Gradstein et al. 2012).

The implementation of RASC in the program PAST version 2.15 was used (Hammer and Harper 2001). For the analysis, sections in rows (one section per row), events in columns (in this case first and last appearances and the stratigraphical level of each horizon in each section is required (in metres or simply ordered), increasing upwards (Hammer and Harper 2006; Hammer 2012). Although the implementation of RASC in PAST is not comprehensive, it comes with some handy features like the ability to automatically transcribe samples to events, i.e. the data input from UA to RASC (Hammer 2012). This eliminates human errors during the data input stage and increases work efficiency. Of the two steps required for this method, the ranking and the scaling, I only carried out the first step. This step produces a single, comprehensive stratigraphical ordering of events. Contradictions are resolved by the "majority vote", i.e. counting the times A occurs above B, or below B or together with B, which is achieved by the Modified Hay Method (Hay 1972; Hammer and Harper 2006; Hammer 2012). Cycles can be resolved by erasing the weakest link (Hammer and Harper 2006). Detailed information about this method and applications can be found in Gradstein et al. (1985) and Hammer (2012), information about the different parameters is given in Hammer (2012) and Hammer and Harper (2006). The default parameters in PAST were used for this analysis.

4.4 Reference sections

CONOP and RASC require reference sections in order to achieve the comparability of the results. The best sampled section is chosen: For the Carboniferous dataset the Oberrödinghausen railway cutting section by Vöhringer (1960) (ORBV) and for the Devonian dataset Müssenberg 1 (M1). I ordered the species according to their FAD and afterwards according to their LAD.

Devonian

In the Devonian Müssenberg 1 (M1) dataset 44 species in 78 horizons occur. There are 44 event horizons (EHs), which represent first or last occurrences (Tab. 9; Fig. 11). These 44 EHs do not coincide with the 78 horizons of the section, because each horizon does not have a FAD or a LAD. The EHs are numbered from 1 (the oldest) to 44 (the youngest). Some of the species occur exclusively at only one of the EHs and hence can be ordered without doubt.

Species	FAD	Horizon	LAD	Horizon	Exclusive
Clymenia laevigata	1	M1-109	2	M1-108	
Kosmoclymenia lamellosa	2	M1-108	2	M1-108	yes
Progonioclymenia acuticostata	2	M1-108	2	M1-108	yes
Cymaclymenia cordata	2	M1-108	10	M1-94	
Kosmoclymenia inaequistriata	3	M1-107	6	M1-101	
Gonioclymenia speciosa	4	M1-104	5	M1-103	
Kosmoclymenia effenbergensis	7	M1-99	8	M1-98	
Mimimitoceras lineare	9	M1-96	13	M1-90	
Piriclymenia piriformis	11	M1-92	12	M1-91	
Rodachia dorsocostata	14	M1-87	14	M1-87	yes
Cymaclymenia warsteinensis	15	M1-86a	15	M1-86a	yes
Cyrtoclymenia plicata	15	M1-86a	15	M1-86a	yes
Cymaclymenia camerata	15	M1-86a	22	M1-53	
Kosmoclymenia undulata	15	M1-86a	25	M1-45	
Muessenbiaergia sublaevis	15	M1-86a	27	M1-33	
Linguaclymenia similis	15	M1-86a	39	M1-15	
Cymaclymenia costellata	16	M1-85	33	M1-25	
Cyrtoclymenia angustiseptata	17	M1-84	29	M1-29	
Mimimitoceras liratum	17	M1-84	32	M1-26	
Cymaclymenia striata	18	M1-83	40	M1-10	
Kalloclymenia subarmata	19	M1-72	31	M1-26a	
Muessenbiaergia bisulcata	20	M1-67	20	M1-67	yes
Muessenbiaergia parundulata	20	M1-67	21	M1-64	
Mimimitoceras geminum	23	M1-40	23	M1-40	yes
Effenbergia lens	24	M1-39	28	M1-31	
Parawocklumeria patens	26	M1-34	27	M1-35	
Glatziella glaucopis	26	M1-34	29	M1-29	
Linguaclymenia clauseni	26	M1-34	34	M1-23b	
					...

Species	FAD	Horizon	LAD	Horizon	Exclusive
...					
Effenbergia falx	27	M1-35	37	M1-15	
Kamptoclymenia endogona	28	M1-31	32	M1-26	
Parawocklumeria paprothae	28	M1-31	32	M1-26	
Kosmoclymenia schindewolfi	28	M1-31	34	M1-23b	
Mimimitoceras trizonatum	28	M1-31	34	M1-23b	
Balvia globulare	28	M1-31	37	M1-15	
Mimimitoceras fuerstenbergi	30	M1-28	30	M1-28	yes
Lissoclymenia wocklumeri	31	M1-26a	43	M1-5	
Mimimitoceras lentum	33	M1-25	38	M1-14	
Parawocklumeria paradoxa	33	M1-25	42	M1-8	
Kenseyoceras nucleus	35	M1-21	43	M1-5	
Kenseyoceras biforme	36	M1-17	42	M1-8	
Postglatziella carinata	37	M1-15	37	M1-15	yes
Finiclymenia wocklumensis	37	M1-15	41	M1-9	
Cymaclymenia involvens	38	M1-14	38	M1-14	yes
Wocklumeria denckmanni	38	M1-14	44	M1-4	

Tab. 9: EHs of the FADs and LADs of the species of the reference section M1.

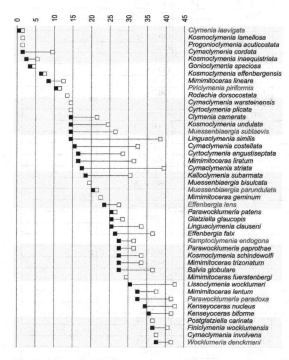

Fig. 11: FADs and LADs (in event horizons) of the species of the reference section M1 (from Klein and Korn 2015).

Carboniferous

In the Carboniferous Oberrödinghausen railway cutting section by Vöhringer (1960) (ORBV), 45 species occur in 10 horizons, which perfectly coincide with the ten EHs (Tab. 10; Fig. 12). Interestingly, almost half of the species (20) have their FAD in the first two horizons (ORBV-6 and ORBV-5).

Species	FAD	Horizon	LAD	Horizon	Exclusive
Acutimitoceras convexum	1	ORBV-6	2	ORBV-5	
Acutimitoceras undulatum	1	ORBV-6	2	ORBV-5	
Weyerella reticulum	1	ORBV-6	2	ORBV-5	
Acutimitoceras acutum	1	ORBV-6	3	ORBV-4	
Acutimitoceras kleinerae	1	ORBV-6	3	ORBV-4	
Acutimitoceras intermedium	1	ORBV-6	4	ORBV-3e	
Gattendorfia subinvoluta	1	ORBV-6	5	ORBV-3d	
Hasselbachia sphaeroidalis	1	ORBV-6	5	ORBV-3d	
Acutimitoceras antecedens	1	ORBV-6	6	ORBV-3c	
Acutimitoceras subbilobatum	1	ORBV-6	10	ORBV-1	
Voehringerites peracutus	2	ORBV-5	2	ORBV-5	yes
Weyerella concava	2	ORBV-5	3	ORBV-4	
Globimitoceras globiforme	2	ORBV-5	4	ORBV-3e	
Mimimitoceras hoennense	2	ORBV-5	5	ORBV-3d	
Nicimitoceras subacre	2	ORBV-5	8	ORBV-3a	
Gattendorfia costata	2	ORBV-5	9	ORBV-2	
Nicimitoceras heterolobatum	2	ORBV-5	9	ORBV-2	
Nicimitoceras trochiforme	2	ORBV-5	9	ORBV-2	
Eocanites nodosus	2	ORBV-5	10	ORBV-1	
Gattendorfia tenuis	2	ORBV-5	10	ORBV-1	
Mimimitoceras varicosum	2	ORBV-5	10	ORBV-1	
Hasselbachia multisulcata	3	ORBV-4	5	ORBV-3d	
Acutimitoceras exile	3	ORBV-4	6	ORBV-3c	
Paprothites dorsoplanus	3	ORBV-4	7	ORBV-3b	
Paragattendorfia globiformis	4	ORBV-3e	10	ORBV-1	
Weyerella molaris	4	ORBV-3e	10	ORBV-1	
Hasselbachia gracilis	5	ORBV-3d	5	ORBV-3d	yes
Paprothites raricostatus	5	ORBV-3d	5	ORBV-3d	yes
Costimitoceras ornatum	5	ORBV-3d	6	ORBV-3c	
Eocanites spiratissimus	6	ORBV-3c	6	ORBV-3c	yes
Eocanites brevis	6	ORBV-3c	7	ORBV-3b	
Eocanites tener	6	ORBV-3c	7	ORBV-3b	
Pseudarietites subtilis	6	ORBV-3c	7	ORBV-3b	
Pseudarietites westfalicus	6	ORBV-3c	8	ORBV-3a	
Eocanites carinatus	7	ORBV-3b	7	ORBV-3b	yes
Acutimitoceras depressum	7	ORBV-3b	9	ORBV-2	
Nicimitoceras acre	7	ORBV-3b	9	ORBV-2	
Acutimitoceras simile	7	ORBV-3b	10	ORBV-1	
Gattendorfia crassa	7	ORBV-3b	10	ORBV-1	
Pseudarietites planissimus	8	ORBV-3a	8	ORBV-3a	yes
Kazakhstania evoluta	9	ORBV-2	9	ORBV-2	yes
Paragattendorfia patens	9	ORBV-2	9	ORBV-2	yes
Eocanites supradevonicus	9	ORBV-2	10	ORBV-1	
Eocanites planus	10	ORBV-1	10	ORBV-1	yes
Paralytoceras serratum	10	ORBV-1	10	ORBV-1	yes

Tab. 10: EHs of the FADs and LADs of the species of the reference section ORBV.

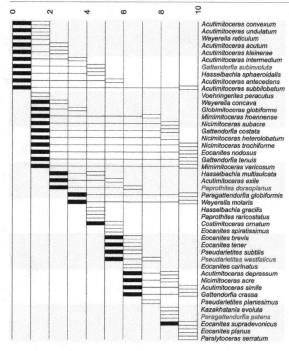

Fig. 12: FADs and LADs (in event horizons) of the species of the reference section ORBV (from Klein and Korn 2015).

5 Results

5.1 Unitary Associations

Devonian

The calculation of the UAs was carried out with different datasets to test which size of dataset and also which modifications lead to the best result (Tab. 11).

Analysis	Dataset	Unitary Associations	Maximal Cliques	Contradictions
A	M1	17	17	0
B	M1 + ORSTA + ORSTB	22	27	60
C	M1 + ORSTA + ORSTB + DASS	21	33	105
D	Complete	22	46	265
E	FAD only	26	43	216
F	Omit singletons	19	38	191
G	Genus level	11	11	0

Tab 11: Overview over the analyses of the Devonian dataset.

Analysis A

Analysed dataset: Müssenberg 1 (M1) (Fig. 13; Tab. 12).

Result in numbers: 17 unitary associations, 17 maximal cliques and 0 contradictions. Maximal cliques are groups of co-occurring taxa, which can not be included in a larger group of co-occurring taxa. With further processing they can be transformed in unitary associations (Hammer 2012). Contradictions result from the incompleteness of the fossil record, which produces conflicting stratigraphical relationships (Monnet et al. 2011). They are resolved using a "majority rule", where the frequency of A below B and B below A are counted and the highest number reflects the right order (Guex 1991; Monnet et al. 2011).

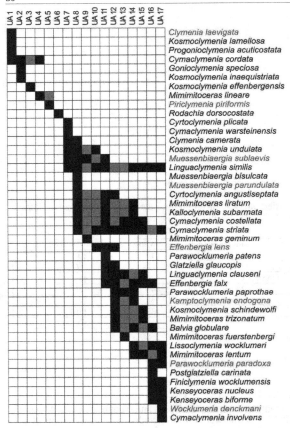

Fig. 13: Result of Analysis A of the Devonian dataset.

UA	Horizons	Defining species	Remarks
UA 1	M1-109 to M1-108	*Clymenia laevigata* (EXC) *Kosmoclymenia lamellosa* (EXC) *Progonioclymenia acuticostata* (EXC) *Cymaclymenia cordata* (FAD)	
UA 2	M1-107 to M1-101	*Gonioclymenia speciosa* (EXC) *Kosmoclymenia inaequistriata* (EXC)	
UA 3	M1-99 to M1-98	*Kosmoclymenia effenbergensis* (EXC)	
UA 4	M1-96 to M1-94	*Mimimitoceras lineare* (FAD) *Cymaclymenia cordata* (LAD)	
UA 5	M1-92 to M1-91	*Piriclymenia piriformis* (EXC)	
			...

UA	Horizons	Defining species	Remarks
...			
UA 6	M1-87	*Rodachia dorsocostata* (EXC)	
UA 7	M1-86a	*Cymaclymenia warsteinensis* (EXC) *Cyrtoclymenia plicata* (EXC) *Cymaclymenia camerata* (FAD) *Kosmoclymenia undulata* (FAD) *Linguaclymenia similis* (FAD) *Muessenbiaergia sublaevis* (FAD)	From UA 7 and higher, the amount of species per horizon increases.
UA 8	M1-85 to M1-53	*Muessenbiaergia bisulcata* (EXC) *Muessenbiaergia parundulata* (EXC) *Cymaclymenia costellata* (FAD) *Cymaclymenia striata* (FAD) *Cyrtoclymenia angustiseptata* (FAD) *Kalloclymenia subarmata* (FAD) *Mimimitoceras liratum* (FAD) *Cymaclymenia camerata* (LAD)	
UA 9	M1-40	*Mimimitoceras geminum* (EXC)	Mainly composed of discontinuities.
UA 10	M1-36 to M1-35b	*Effenbergia lens* (FAD) *Kosmoclymenia undulata* (LAD)	
UA 11	M1-34 to M1-33	*Parawocklumeria patens* (EXC) *Linguaclymenia clauseni* (FAD) *Muessenbiaergia sublaevis* (LAD)	
UA 12	M1-31 to M1-29	*Balvia globulare* (FAD) *Kamptoclymenia endogona* (FAD) *Kosmoclymenia schindewolfi* (FAD) *Mimimitoceras trizonatum* (FAD) *Parawocklumeria paprothae* (FAD) *Cyrtoclymenia angustiseptata* (LAD) *Effenbergia lens* (LAD) *Glatziella glaucopis* (LAD)	
UA 13	M1-28	*Mimimitoceras fuerstenbergi* (EXC)	
UA 14	M1-26a to M1-26	*Lissoclymenia wocklumeri* (FAD) *Kalloclymenia subarmata* (LAD) *Kamptoclymenia endogona* (LAD) *Parawocklumeria paprothae* (LAD) *Mimimitoceras liratum* (LAD)	
UA 15	M1-25 to M1-23	*Mimimitoceras lentum* (FAD) *Parawocklumeria paradoxa* (FAD) *Cymaclymenia costellata* (LAD) *Linguaclymenia clauseni* (LAD) *Kosmoclymenia schindewolfi* (LAD) *Mimimitoceras trizonatum* (LAD)	
UA 16	M1-21 to M1-15	*Postglatziella carinata* (EXC) *Finiclymenia wocklumensis* (FAD) *Kenseyoceras biforme* (FAD) *Kenseyoceras nucleus* (FAD) *Balvia globulare* (LAD) *Effenbergia falx* (LAD)	
			...

UA	Horizons	Defining species	Remarks
...			
UA 17	M1-14 to M1-4	*Cymaclymenia involvens* (EXC) *Wocklumeria denckmanni* (EXC)	

Tab. 12: Description of the unitary associations obtained by Analysis A of the Devonian dataset, including the horizons, which can be assigned to the UAs. Species can define a UA by (1) an exclusive occurrence (EXC), (2) a first appearance (FAD) or (3) a last appearance (LAD).

Reliability – Analysis A includes many uncertainties and discontinuities but no contradictions occur, 67 of the 78 horizons can be clearly assigned to UAs.

Analysis B

Analysed dataset: Müssenberg 1 (M1) + Oberrödinghausen road cutting alpha (ORSTA) and Oberrödinghausen road cutting beta (ORSTB) (Fig. 14; Tab. 13).

Result in numbers: 22 unitary associations, 27 maximal cliques and 60 contradictions.

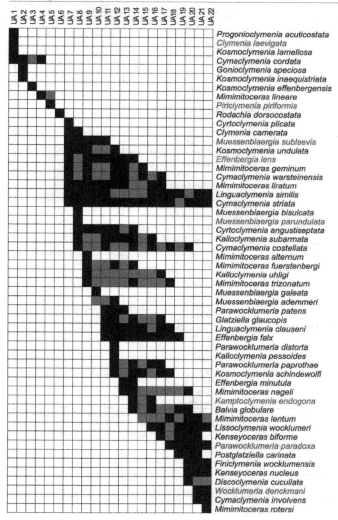

Fig. 14: Result of Analysis B of the Devonian dataset.

UA	Horizons	Defining species	Remarks
The UAs 1 to 6 are identical to the UAs 1 to 6 in the analysis of the section M1.			
			...

UA	Horizons	Defining species	Remarks
...			
UA 7	ORSTB-12(3) to ORSTB-12(2) M1-86a	*Cyrtoclymenia plicata* (EXC) *Cymaclymenia camerata* (FAD) *Cymaclymenia striata* (FAD) *Cymaclymenia warsteinensis* (FAD) *Effenbergia lens* (FAD) *Kosmoclymenia undulata* (FAD) *Linguaclymenia similis* (FAD) *Mimimitoceras geminum* (FAD) *Mimimitoceras liratum* (FAD) *Muessenbiaergia sublaevis* (FAD)	The only association, which is shared by M1 and ORSTB.
UA 8	M1-85 to M1-53	*Muessenbiaergia bisulcata* (EXC) *Muessenbiaergia parundulata* (EXC) *Cymaclymenia costellata* (FAD) *Cyrtoclymenia angustiseptata* (FAD) *Kalloclymenia subarmata* (FAD) *Cymaclymenia camerata* (LAD)	
UA 9	ORSTB-11(1)	*Mimimitoceras alternum* (EXC) *Kalloclymenia uhligi* (FAD) *Mimimitoceras fuerstenbergi* (FAD) *Mimimitoceras trizonatum* (FAD)	
UA 10	ORSTB-12(1)	*Muessenbiaergia galeata* (EXC)	
UA 11	M1-34 to M1-33	*Effenbergia falx* (FAD) *Glatziella glaucopis* (FAD) *Linguaclymenia clauseni* (FAD) *Parawocklumeria patens* (FAD) *Muessenbiaergia sublaevis* (LAD)	
UA 12	ORSTA-9a(3) ORSTB-9b(2) to ORSTB-8b(1+2)	*Kalloclymenia pessoides* (EXC) *Parawocklumeria distorta* (EXC) *Kosmoclymenia undulata* (LAD) *Muessenbiaergia ademmeri* (LAD) *Kosmoclymenia schindewolfi* (FAD) *Parawocklumeria paprothae* (FAD)	
UA 13	ORSTB-8b(1) to ORSTB-7b(4)	*Effenbergia minutula* (FAD) *Mimimitoceras nageli* (FAD) *Parawocklumeria patens* (LAD)	
UA 14	ORSTA-8a(2) M1-31 to M1-28	*Balvia globulare* (FAD) *Kamptoclymenia endogona* (FAD) *Cyrtoclymenia angustiseptata* (LAD) *Effenbergia lens* (LAD) *Mimimitoceras fuerstenbergi* (LAD)	Shared by M1 and ORSTA.
UA 15	ORSTA-7a(2)	*Mimimitoceras geminum* (LAD)	
UA 16	M1-26a to M1-26	*Lissoclymenia wocklumeri* (LAD) *Kalloclymenia subarmata* (LAD)	
UA 17	ORSTA-7a(1) to ORSTA-6a(2+3) ORSTB-6b(3)	*Kenseyoceras biforme* (FAD) *Cymaclymenia warsteinensis* (LAD) *Glatziella glaucopis* (LAD) *Kalloclymenia uhligi* (LAD) *Mimimitoceras liratum* (LAD) *Parawocklumeria paprothae* (LAD)	
			...

UA	Horizons	Defining species	Remarks
...			
UA 18	M1-25 to M1-23b	*Parawocklumeria paradoxa* (FAD) *Linguaclymenia clauseni* (LAD) *Kosmoclymenia schindewolfi* (LAD) *Mimimitoceras trizonatum* (LAD)	
UA 19	M1-21 to M1-15	*Postglatziella carinata* (FAD) *Finiclymenia wocklumensis* (FAD) *Kenseyoceras nucleus* (FAD) *Effenbergia falx* (LAD)	
UA 20	ORSTA-4(3) to ORSTA-3(4)	*Discoclymenia cucullata* (FAD) *Cymaclymenia costellata* (LAD) *Mimimitoceras nageli* (LAD)	
UA 21	ORSTA-3(3) to ORSTA-3(2) M1-14 to M1-8	*Cymaclymenia involvens* (FAD) *Wocklumeria denckmanni* (FAD)	Shared by M1 and ORSTA.
UA 22	ORSTA-2(1)	*Mimimitoceras rotersi* (EXC)	

Tab. 13: Description of the unitary associations obtained by Analysis B of the Devonian dataset, including the horizons, which can be assigned to the UAs. Species can define a UA by (1) an exclusive occurrence (EXC), (2) a first appearance (FAD) or (3) a last appearance (LAD).

Reliability – 85 of the 133 horizons can be clearly assigned to UAs.

Comparison of Analysis A and Analysis B – Differences are: (1) *Mimimitoceras geminum* and *Effenbergia lens*, which form UA 9 and 10 in Analysis A, are grouped to UA 7 in Analysis B, (2) *Mimimitoceras fuerstenbergi*, which forms UA 13 in Analysis A, forms UA 9 in Analysis B and (3) the UAs 14 to 16 of Analysis A are not clearly resolved in Analysis B.

Analysis C

Analysed dataset: Müssenberg 1 (M1) + Oberrödinghausen road cutting alpha (ORSTA) + Oberrödinghausen road cutting beta (ORSTB) + Dasberg South (DASS) (Fig. 15; Tab. 14).

Result in numbers: 21 unitary associations, 33 maximal cliques, 105 contradictions and 3 cliques in cycles.

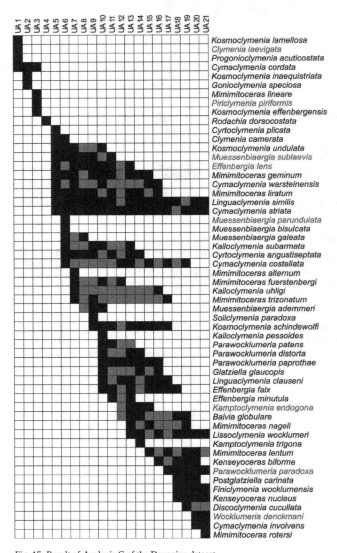

Fig. 15: Result of Analysis C of the Devonian dataset.

UA	Horizons	Defining species	Remarks
UA 1	M1-109 to M1-108	*Clymenia laevigata* (EXC) *Kosmoclymenia lamellosa* (EXC) *Progonioclymenia acuticostata* (EXC) *Cymaclymenia cordata* (FAD)	
UA 2	M1-107 to M1-101 DASS-55 to DASS-51	*Gonioclymenia speciosa* (EXC) *Kosmoclymenia inaequistriata* (EXC)	
UA 3	M1-99 to M1-90 DASS-46	*Kosmoclymenia effenbergensis* (EXC) *Mimimitoceras lineare* (EXC) *Piriclymenia piriformis* (EXC)	
UA 4	M1-87	*Rodachia dorsocostata* (EXC)	
UA 5	ORSTB-12(3) to ORSTB-12(2) M1-87 to M1-86a	*Cymaclymenia camerata* (FAD) *Cymaclymenia striata* (FAD) *Cymaclymenia warsteinensis* (FAD) *Effenbergia lens* (FAD) *Kosmoclymenia undulata* (FAD) *Linguaclymenia similis* (FAD) *Mimimitoceras geminum* (FAD) *Mimimitoceras liratum* (FAD) *Muessenbiaergia sublaevis* (FAD)	
UA 6	M1-85 to M1-53 DASS-35 to DASS19	*Muessenbiaergia bisulcata* (EXC) *Muessenbiaergia parundulata* (EXC) *Cymaclymenia costellata* (FAD) *Cyrtoclymenia angustiseptata* (FAD) *Kalloclymenia subarmata* (FAD) *Muessenbiaergia galeata* (FAD) *Cymaclymenia camerata* (LAD)	Represented by the most horizons, but most of them only contain a small amount of species, some of them only one.
UA 7	ORSTB-12(1)	*Mimimitoceras alternum* (EXC) *Kalloclymenia uhligi* (FAD) *Mimimitoceras fuerstenbergi* (FAD) *Mimimitoceras trizonatum* (FAD)	
UA 8	ORSTB-11(1)	*Muessenbiaergia galeata* (LAD)	
UA 9	DASS-8 to DASS-7	*Soliclymenia paradoxa* (EXC) *Kosmoclymenia schindewolfi* (FAD)	
UA 10	ORSTB-10(1) to ORSTB-8b(2)	*Kalloclymenia pessoides* (EXC) *Effenbergia falx* (FAD) *Glatziella glaucopis* (FAD) *Linguaclymenia clauseni* (FAD) *Parawocklumeria distorta* (FAD) *Parawocklumeria paprothae* (FAD) *Parawocklumeria patens* (FAD) *Kosmoclymenia undulata* (LAD) *Muessenbiaergia ademmeri* (LAD)	
UA 11	DASS-5 to DASS-3	*Effenbergia minutula* (FAD) *Muessenbiaergia sublaevis* (LAD)	
UA 12	M1-28	*Mimimitoceras fuerstenbergi* (LAD)	
UA 13	ORSTA-8a(2) ORSTB-7b(5) M1-26a DASS-2	*Lissoclymenia wocklumeri* (FAD) *Mimimitoceras nageli* (FAD) *Effenbergia lens* (LAD) *Kalloclymenia subarmata* (LAD)	The only association, which shares the combined information of all sections.
UA 14	DASS-1	*Kamptoclymenia trigona* (EXC) *Cyrtoclymenia angustiseptata* (LAD) *Parawocklumeria distorta* (LAD)	
			...

UA	Horizons	Defining species	Remarks
...			
UA 15	ORSTA-7a(3) to ORSTA-7a(2)	*Mimimitoceras lentum* (FAD) *Kamptoclymenia endogona* (LAD) *Mimimitoceras geminum* (LAD)	
From UA 16 to 21 the succession is equivalent to the UAs 17 to 22 in Analysis B. Only horizons from M1 and ORSTA are contained.			

Tab. 14: Description of the unitary associations obtained by Analysis C of the Devonian dataset, including the horizons, which can be assigned to the UAs. Species can define a UA by (1) an exclusive occurrence (EXC), (2) a first appearance (FAD) or (3) a last appearance (LAD).

Reliability – 104 of the 174 horizons can be clearly assigned to UAs.

Comparison of Analysis A and Analysis C – The most noticeable differences between the results of Analysis A and Analysis C are: (1) *Mimimitoceras geminum* and *Effenbergia lens* are grouped to UA 5 in Analysis C, (2) *Mimimitoceras fuerstenbergi* and *Mimimitoceras trizonatum* form UA 7 in Analysis C, (3) the UAs 11 and 12 of Analysis A form the UAs 9 to 12 in Analysis C and (4) the UAs 14 to 16 of Analysis A are not clearly resolved in Analysis C.

Analysis D

Analysed dataset: The complete corrected Devonian dataset, which contains thirteen sections (Fig. 16; Tab. 15).

Result in numbers: 22 unitary associations, 46 maximal cliques, 265 contradictions and 1 residual virtual edge. Edges represent compatible species, when they represent virtually coexisting species they are called "virtual" edges. Those species are never included in maximal horizons, from which the maximal cliques are calculated (Guex 1991).

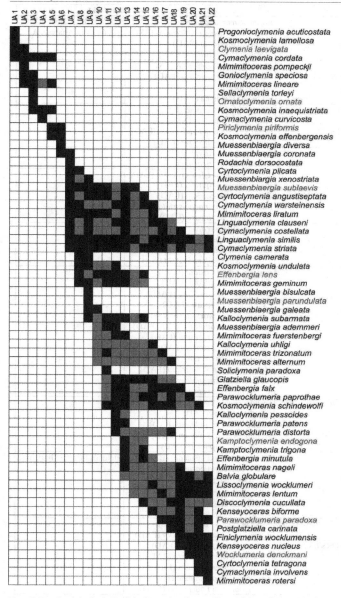

UA 1	UA 2	UA 3	UA 4	UA 5	UA 6	UA 7	UA 8	UA 9	UA 10	UA 11	UA 12	UA 13	UA 14	UA 15	UA 16	UA 17	UA18	UA 19	UA 20	UA 21	UA 22	

Progonioclymenia acuticostata
Kosmoclymenia lamellosa
Clymenia laevigata
Cymaclymenia cordata
Mimimitoceras pompeckji
Gonioclymenia speciosa
Mimimitoceras lineare
Sellaclymenia torleyi
Ornatoclymenia ornata
Kosmoclymenia inaequistriata
Cymaclymenia curvicosta
Piriclymenia piriformis
Kosmoclymenia effenbergensis
Muessenbiaergia diversa
Muessenbiaergia coronata
Rodachia dorsocostata
Cyrtoclymenia plicata
Muessenbiargia xenostriata
Muessenbiaergia sublaevis
Cyrtoclymenia angustiseptata
Cymaclymenia warsteinensis
Mimimitoceras liratum
Linguaclymenia clauseni
Cymaclymenia costellata
Linguaclymenia similis
Cymaclymenia striata
Clymenia camerata
Kosmoclymenia undulata
Effenbergia lens
Mimimitoceras geminum
Muessenbiaergia bisulcata
Muessenbiaergia parundulata
Muessenbiaergia galeata
Kalloclymenia subarmata
Muessenbiaergia ademmeri
Mimimitoceras fuerstenbergi
Kalloclymenia uhligi
Mimimitoceras trizonatum
Mimimitoceras alternum
Soliclymenia paradoxa
Glatziella glaucopis
Effenbergia falx
Parawocklumeria paprothae
Kosmoclymenia schindewolfi
Kalloclymenia pessoides
Parawocklumeria patens
Parawocklumeria distorta
Kamptoclymenia endogona
Kamptoclymenia trigona
Effenbergia minutula
Mimimitoceras nageli
Balvia globulare
Lissoclymenia wocklumeri
Mimimitoceras lentum
Discoclymenia cucullata
Kenseyoceras biforme
Parawocklumeria paradoxa
Postglatziella carinata
Finiclymenia wocklumensis
Kenseyoceras nucleus
Wocklumeria denckmani
Cyrtoclymenia tetragona
Cymaclymenia involvens
Mimimitoceras rotersi

Fig. 16: Result of Analysis D of the Devonian dataset (from Klein and Korn 2015).

UA	Horizons	Defining species	Remarks
UA 1	M1-108	*Kosmoclymenia lamellosa* (EXC), *Progonioclymenia acuticostata* (EXC) *Clymenia laevigata* (FAD) *Cymaclymenia cordata* (FAD)	Includes the same four species as in Analysis A.
UA 2	E77-B DASN-18 to DASN-14	*Mimimitoceras pompeckji* (EXC) *Gonioclymenia speciosa* (FAD) *Mimimitoceras lineare* (FAD) Clymenia laevigata (LAD)	
UA 3	M1-104 to M1-103 M3-6 to M3-4a DASN-19 to DASN-22	*Ornatoclymenia ornata* (EXC) *Sellaclymenia torleyi* (EXC) *Kosmoclymenia inaequistriata* (FAD) *Gonioclymenia speciosa* (LAD)	
UA 4	E77-G	*Cymaclymenia curvicosta* (EXC) *Cymaclymenia tricarinata* (EXC)	
UA 5	M1-99 to M1-91 M3-3 to M3-1	*Kosmoclymenia effenbergensis* (FAD) *Piriclymenia piriformis* (FAD) *Cymaclymenia cordata* (LAD) *Kosmoclymenia inaequistriata* (LAD) *Mimimitoceras lineare* (LAD)	
UA 6	ORSK-20 E77-J	*Muessenbiaergia diversa* (EXC) *Muessenbiaergia coronata* (FAD) *Kosmoclymenia effenbergensis* (LAD) *Piriclymenia piriformis* (LAD)	
UA 7	E77-L to E77-M M1-87	*Rodachia dorsocostata* (EXC) *Cyrtoclymenia plicata* (FAD) *Cymaclymenia costellata* (FAD) *Cymaclymenia striata* (FAD) *Cymaclymenia warsteinensis* (FAD) *Cyrtoclymenia angustiseptata* (FAD) *Linguaclymenia clauseni* (FAD) *Linguaclymenia similis* (FAD) *Mimimitoceras liratum* (FAD) *Muessenbiargia xenostriata* (FAD) *Muessenbiaergia sublaevis* (FAD) *Muessenbiaergia coronata* (LAD)	
UA 8	ORSTB-12(3) to ORSTB-12(2) M1-86a	*Cymaclymenia camerata* (FAD) *Effenbergia lens* (FAD) *Kosmoclymenia undulata* (FAD) *Mimimitoceras geminum* (FAD) *Cyrtoclymenia plicata* (LAD)	
UA 9	E77-N to E77-Q E87-N2 to E87-O3 M1-72 to M1-53 DASS-35 toDASS-19	*Muessenbiaergia bisulcata* (EXC) *Muessenbiaergia parundulata* (EXC) *Kalloclymenia subarmata* (FAD) *Muessenbiaergia galeata* (FAD) *Muessenbiargia xenostriata* (LAD)	Includes the most horizons.
UA 10	ORSTB-11(1)	*Muessenbiaergia galeata* (LAD)	
UA 11	DASS-8 to DASS-7	*Soliclymenia paradoxa* (EXC) *Kosmoclymenia schindewolfi* (FAD)	
UA 12	ORSTB-9b(2) to ORSTB-8b(2)	*Kalloclymenia pessoides* (FAD) *Parawocklumeria distorta* (FAD) *Parawocklumeria patens* (FAD) *Kosmoclymenia undulata* (LAD) *Muessenbiaergia ademmeri* (LAD)	
			...

UA	Horizons	Defining species	Remarks
...			
UA 13	ORSK-14 to ORSK-10 ORSTA-9a(1) ORSTB-8b(1) to ORSTB-7b(4) M1-28 DASS-4	*Effenbergia minutula* (FAD) *Kamptoclymenia endogona* (FAD) *Kamptoclymenia trigona* (FAD) *Mimimitoceras nageli* (FAD) *Mimimitoceras fuerstenbergi* (LAD) *Parawocklumeria patens* (LAD)	
UA 14	ORBK-10B	*Muessenbiaergia sublaevis* (LAD)	Mainly composed of discontinuities.
UA 15	M4B-6	*Discoclymenia cucullata* (FAD) *Cyrtoclymenia angustiseptata* (LAD) *Effenbergia lens* (LAD) *Kalloclymenia subarmata* (LAD)	
UA 16	ORSTA-7a(1) to ORSTA-6a(2+3) ORSTB-6b(3)	*Kenseyoceras biforme* (FAD) *Cymaclymenia warsteinensis* (LAD) *Kalloclymenia uhligi* (LAD) *Mimimitoceras liratum* (LAD)	
UA 17	M1-25 to M1-23b	*Parawocklumeria paradoxa* (FAD) *Mimimitoceras trizonatum* (LAD)	
UA 18	ORBK-6 to ORBK-5	*Postglatziella carinata* (FAD) *Mimimitoceras alternum* (LAD) *Parawocklumeria distorta* (LAD)	
UA 19	ORSTA-5(2) to ORSTA-3(4) M1-21 to M1-15	*Finiclymenia wocklumensis* (FAD) *Kenseyoceras nucleus* (FAD) *Cymaclymenia costellata* (LAD) *Effenbergia falx* (LAD)	
UA 20	M4A-8 to M4A-5	*Wocklumeria denckmanni* (FAD) *Parawocklumeria paprothae* (LAD)	
UA 21	ORSK-5 to ORSK-4 ORSTA-3(3) to ORSTA-3(2) M1-14 DASM-3	*Cymaclymenia involvens* (FAD) *Cyrtoclymenia tetragona* (FAD) *Kosmoclymenia schindewolfi* (LAD)	
UA 22	ORSK-1 ORSTA-2(1)	*Mimimitoceras rotersi* (EXC)	

Tab. 15: Description of the unitary associations obtained by Analysis D of the Devonian dataset, including the horizons, which can be assigned to the UAs. Species can define a UA by (1) an exclusive occurrence (EXC), (2) a first appearance (FAD) or (3) a last appearance (LAD).

Reliability – Only 112 of the 283 horizons can be clearly assigned to UAs. This is the result of the large amount of contradictions as well as the residual virtual edge.

Stratigraphical order of the sections – M1 ranges almost through the complete succession from UA 1 to UA 21. ORSK almost covers the complete range, too, but only the UAs 6, 13, 21 and 22 are present. DASN (up to UA 3) and M3 (up to UA 5) cover the lower part of the complete succession, E 77 covers UA 2 to 9, E 87 is only present in UA 9. The central part of the succession is provided by DASS (UA 9 to 13) and ORSTB (UA 8 to 16). The two UAs 14 and 18 from ORBK and 15 and 20 from M4 constitute to the upper part of the succession, DASM only constitutes of UA 21. ORSTA covers the upper part of the succession form UA 13 to UA 22. DD does not provide a safe horizon and thus is hard to put in order without doubt.

Comparison of Analysis A and Analysis D – Differences are: (1) the UAs 7 and 8 of Analysis A can not be distinguished in Analysis D, (2) *Linguaclymenia clauseni* has its FAD in UA 11 in Analysis A but in UA 7 in Analysis D, (3) the UAs 11 and 12 of Analysis A form the UAs

10 to 13 in Analysis D and (4) the UAs 14 to 16 of Analysis A are not clearly resolved in Analysis D. In conclusion, the congruence of Analysis D with the reference section can be considered as better than in Analysis B and C and the resolution is higher.

Analysis E

Analysed dataset: The complete corrected Devonian dataset with first occurrences only (Fig. 17; Tab. 16).

Result in numbers: 26 unitary associations, 43 maximal cliques, 213 contradictions and 11 residual virtual edges.

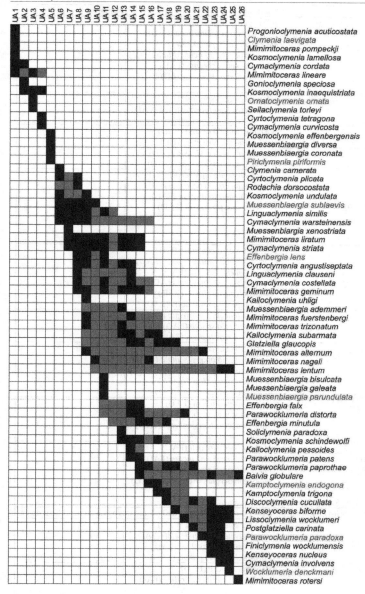

Fig. 17: Result of Analysis E of the Devonian dataset.

UA	Horizons	Defining species	Remarks
UA 1	E77-B M1-109 to M1-108 M3-7 DASN-9	*Clymenia laevigata* (FAD) *Kosmoclymenia lamellosa* (FAD) *Mimimitoceras pompeckji* (FAD) *Progonioclymenia acuticostata* (FAD) *Cymaclymenia cordata* (FAD) *Mimimitoceras lineare* (FAD)	
UA 2	M1-104 M3-6 DASN-14	*Gonioclymenia speciosa* (FAD) *Kosmoclymenia inaequistriata* (FAD)	
UA 3	M3-4a DASN-19	*Ornatoclymenia ornata* (FAD) *Sellaclymenia torleyi* (FAD)	
UA 4	E77-G	*Cymaclymenia curvicosta* (FAD) *Cymaclymenia tricarinata* (FAD)	
UA 5	ORSK-20 E77-J E87-J4 M1-99 to M1-92 M3-3 to M3-1 DASN-26 DASS-46	*Kosmoclymenia effenbergensis* (FAD) *Muessenbiaergia coronata* (FAD) *Muessenbiaergia diversa* (FAD) *Piriclymenia piriformis* (FAD)	
UA 6	M-86a	*Cymaclymenia camerata* (FAD) *Cymaclymenia warsteinensis* (FAD) *Cyrtoclymenia plicata* (FAD) *Kosmoclymenia undulata* (FAD) *Linguaclymenia similis* (FAD) *Muessenbiaergia sublaevis* (FAD)	
UA 7	E77-L	*Muessenbiargia xenostriata* (FAD) *Cymaclymenia striata* (FAD) *Mimimitoceras liratum* (FAD)	
UA 8	ORSTB-12(3) E77-M	*Cymaclymenia costellata* (FAD) *Cyrtoclymenia angustiseptata* (FAD) *Effenbergia lens* (FAD) *Linguaclymenia clauseni* (FAD) *Mimimitoceras geminum* (FAD)	
UA 9	ORSTB-12(1)	*Kalloclymenia uhligi* (FAD) *Mimimitoceras alternum* (FAD) *Mimimitoceras fuerstenbergi* (FAD) *Mimimitoceras trizonatum* (FAD)	
UA 10	ORSK-14	*Mimimitoceras nageli* (FAD)	
UA 11	ORSTB-11(1) E77-N to E77-Q E87-N2 M1-67 DASS-39 toDASS-29	*Muessenbiaergia bisulcata* (FAD) *Muessenbiaergia galeata* (FAD) *Muessenbiaergia parundulata* (FAD)	
UA 12	Although 17 species contribute to UA 12, no horizon can be assigned to this association without doubt.		
UA 13	DASS-8	*Soliclymenia paradoxa* (FAD) *Kosmoclymenia schindewolfi* (FAD)	
UA 14	ORSK-12 M1-34	*Kalloclymenia pessoides* (FAD) *Parawocklumeria patens* (FAD)	
UA 15	ORSTB-(9b(2) DASM-5 DASS-4	*Balvia globulare* (FAD) *Parawocklumeria paprothae* (FAD)	
			...

UA	Horizons	Defining species	Remarks
...			
UA 16	ORSTA-8a(2)	*Kamptoclymenia endogona* (FAD)	The following UAs are formed by single horizons.
UA 17	ORSK-11	*Kamptoclymenia trigona* (FAD)	
UA 18	M4B-6	*Discoclymenia cucullata* (FAD)	
UA 19	ORSTA-7a(1)	*Kenseyoceras biforme* (FAD)	
UA 20	DASS-2	*Lissoclymenia wocklumeri* (FAD)	
UA 21	ORBK-6	*Postglatziella carinata* (FAD)	
UA 22	ORBK-5	*Parawocklumeria paradoxa* (FAD)	
UA 23	ORSTA-4(3)	*Finiclymenia wocklumensis* (FAD) *Kenseyoceras nucleus* (FAD)	
UA 24	ORSK-4 DASM-3	*Cymaclymenia involvens* (FAD)	
UA 25	ORBK-4 ORSK-2 ORSTA-3(3) M1-14 M4A-8 DASM-1-b DD-90	*Wocklumeria denckmanni* (FAD)	Is formed by horizons from seven sections.
UA 26	ORSK-1 ORSTA-2(1)	*Mimimitoceras rotersi* (FAD)	

Tab. 16: Description of the unitary associations obtained by Analysis E of the Devonian dataset, including the horizons, which can be assigned to the UAs. Species can define a UA by (1) an exclusive occurrence (EXC) or (2) a first appearance (FAD).

Reliability – Only 59 of the 283 horizons can be clearly assigned to UAs.

Comparison of Analysis D and Analysis E – *Linguaclymenia clauseni* is ordered differently, but still has the same co-occurrences. *Mimimitoceras lentum* is not grouped with *Lissoclymenia wocklumeri* and *Parawocklumeria paradoxa*, but they still occur in the same UAs. *Muessenbiaergia bisulcata* and *Muessenbiaergia parundulata* are also ordered differently, but the co-occurrences are mainly the same. New co-occurrences are added by discontinuities (e.g. *Cymaclymenia warsteinensis* and *Mimimitoceras geminum*). The only taxon, which shows a different fossil association is *Mimimitoceras fuerstenbergi*. The taxa, which co-occur with this species in Analysis D only co-occurr via discontinuities in Analysis E. In conclusion, the content and grouping of associations differs from Analysis D a lot, this analysis is not suitable for further investigations.

Analysis F

Analysed dataset: The complete corrected Devonian dataset with singletons excluded (Fig. 18; Tab. 17).

Result in numbers: 19 unitary associations, 38 maximal cliques, 191 contradictions and 1 residual virtual edge.

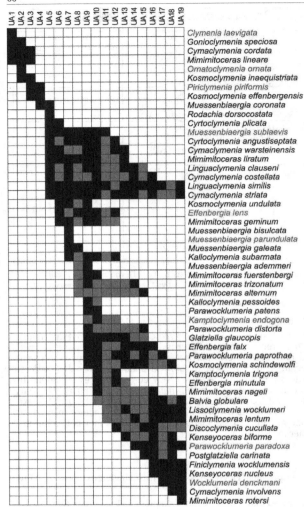

Fig. 18: Result of Analysis F of the Devonian dataset.

UA	Horizons	Defining species	Remarks
UA 1	E77-B M1-109 to M1-108 M3-7 DASN-18 to DASN-9	*Clymenia laevigata* (EXC) *Cymaclymenia cordata* (FAD) *Gonioclymenia speciosa* (FAD) *Mimimitoceras lineare* (FAD)	
UA 2	M1-104 to M1-103 M3-6 to M3-4a DASN-22 to DASN-19	*Ornatoclymenia ornata* (EXC) *Kosmoclymenia inaequistriata* (FAD) *Gonioclymenia speciosa* (LAD)	
UA 3	M1-99 to M1-91 M3-3 to M3-1	*Kosmoclymenia effenbergensis* (FAD) *Piriclymenia piriformis* (FAD) *Cymaclymenia cordata* (LAD) *Kosmoclymenia inaequistriata* (LAD) *Mimimitoceras lineare* (LAD)	
UA 4	ORSK-20 E77-J	*Muessenbiaergia coronata* (FAD) *Kosmoclymenia effenbergensis* (LAD) *Piriclymenia piriformis* (LAD)	
UA 5	E77-L to E77-M M1-87	*Rodachia dorsocostata* (EXC) *Cyrtoclymenia plicata* (FAD) *Cymaclymenia costellata* (FAD) *Cymaclymenia striata* (FAD) *Cymaclymenia warsteinensis* (FAD) *Cyrtoclymenia angustiseptata* (FAD) *Linguaclymenia clauseni* (FAD) *Linguaclymenia similis* (FAD) *Mimimitoceras liratum* (FAD) *Muessenbiaergia sublaevis* (FAD) *Muessenbiaergia coronata* (LAD)	
UA 6	ORSTB-12(3) to ORSTB-12(2) M1-86a	*Effenbergia lens* (FAD) *Kosmoclymenia undulata* (FAD) *Mimimitoceras geminum* (FAD) *Cyrtoclymenia plicata* (LAD)	
UA 7	E77-N to E77-Q E87-N2 to E87-O3 M1-67 to M1-64 DASS-35 to DASS-19	*Muessenbiaergia bisulcata* (EXC) *Muessenbiaergia parundulata* (EXC) *Kalloclymenia subarmata* (FAD) *Muessenbiaergia galeata* (FAD)	
UA 8	ORSTB-11(1)	*Muessenbiaergia galeata* (LAD)	
UA 9	ORSTB-10(1) to ORSTB-8b(2) DASS-8 to DASS-7	*Effenbergia falx* (FAD) *Glatziella glaucopis* (FAD) *Kalloclymenia pessoides* (FAD) *Kosmoclymenia schindewolfi* (FAD) *Parawocklumeria distorta* (FAD) Parawocklumeria paprothae (FAD) *Parawocklumeria patens* (FAD)	
UA 10	ORSK-14 to ORSK-10 ORSTA-9a(1) ORSTB-8b(1) to ORSTB-7b(4) M1-28 DASS-4	*Effenbergia minutula* (FAD) *Kamptoclymenia trigona* (FAD) *Mimimitoceras nageli* (FAD) *Mimimitoceras fuerstenbergi* (LAD) *Parawocklumeria patens* (LAD)	
			...

UA	Horizons	Defining species	Remarks
...			
UA 11	ORBK-10B	*Muessenbiaergia sublaevis* (LAD)	Mainly composed of discontinuities. Contains only three taxa.
UA 12	M4B-6	*Discoclymenia cucullata* (FAD) *Cyrtoclymenia angustiseptata* (LAD) *Effenbergia lens* (LAD) *Kalloclymenia subarmata* (LAD) *Kamptoclymenia trigona* (LAD)	
UA 13	ORSTA-6a(2+3) ORSTA-7a(1) ORSTB-6b(3)	*Kenseyoceras biforme* (FAD) *Cymaclymenia warsteinensis* (LAD) *Mimimitoceras liratum* (LAD)	
UA 14	M1-25 to M1-23b	*Parawocklumeria paradoxa* (FAD) *Mimimitoceras trizonatum* (LAD)	Same as UA 17 in Analysis D.
UA 15	ORBK-6 to ORBK-5	*Postglatziella carinata* (FAD) *Mimimitoceras alternum* (LAD) *Parawocklumeria distorta* (LAD)	Same as UA 18 in Analysis D.
UA 16	ORSTA-5(2) to ORSTA-3(4) M1-21 to M1-15	*Finiclymenia wocklumensis* (FAD) *Kenseyoceras nucleus* (FAD) *Cymaclymenia costellata* (LAD) *Effenbergia falx* (LAD)	Same as UA 19 in Analysis D.
UA 17	M4A-8 to M4A-5	*Wocklumeria denckmanni* (FAD) *Parawocklumeria paprothae* (LAD)	Same as UA 20 in Analysis D.
UA 18	ORSK-5 to ORSK-4 ORSTA-3(3) to ORSTA-3(2) M1-14 DASM-3	*Cymaclymenia involvens* (FAD) *Kosmoclymenia schindewolfi* (LAD)	
UA 19	ORSK-1 ORSTA-2(1)	*Mimimitoceras rotersi* (EXC)	Same as UA 22 in Analysis D.

Tab. 17: Description of the unitary associations obtained by Analysis F of the Devonian dataset, including the horizons, which can be assigned to the UAs. Species can define a UA by (1) an exclusive occurrence (EXC), (2) a first appearance (FAD) or (3) a last appearance (LAD).

Reliability – 12 species are represented by singletons and hence not considered in this analysis. Only 108 of the 283 horizons can be clearly assigned to UAs.

Comparison of Analysis D and Analysis F – The great majority of the UAs are the same in both analyses. The reduced number of UAs occurs due to the missing singletons, e.g. *Progonioclymenia acuticostata* and *Kosmoclymenia lamellosa*, which form UA 1 in Analysis D. Since the majority of the UAs are the same, the succession of the sections is very similar, too. The succession and content of the UAs does not vary much from the reference section. The range of *Linguaclymenia clauseni* is extended due to co-occurrences. *Mimimitoceras trizonatum* and *Mimimitoceras fuerstenbergi* do not co-occur in Analysis F but in the reference section.

Analysis G

Analysed dataset: The complete corrected Devonian dataset on genus level (Fig. 19; Tab. 18).

Result in numbers: 11 unitary associations, 11 maximal cliques, 0 contradictions.

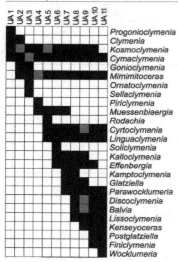

Fig. 19: Result of Analysis G of the Devonian dataset.

UA	Horizons	Defining species	Remarks
UA 1	M1-108	*Progonioclymenia* (EXC) *Cymaclymenia* (FAD) *Clymenia* (FAD) *Kosmoclymenia* (FAD)	
UA 2	E77-B DASN-18 to DASN-14	*Gonioclymenia* (EXC) *Mimimitoceras* (FAD) *Clymenia* (LAD)	
UA 3	M3-4a DASN-21 to DASN-19	*Ornatoclymenia* (EXC) *Sellaclymenia* (EXC)	
UA 4	E77-J E87-J4 M1-92 --M1-91 M3-1 DASS-46	*Piriclymenia* (EXC) *Muessenbiaergia* (FAD)	
UA 5	E77-M M1-87	*Rodachia* (EXC) *Cyrtoclymenia* (FAD) *Linguaclymenia* (FAD)	
UA 6	DASS-8 to DASS-7	*Soliclymenia* (EXC) *Effenbergia* (FAD) *Kalloclymenia* (FAD)	
UA 7	ORBK-10B ORSK-12 to ORSK-9 ORSTB-9b(2) to ORSTB-8b(2) M1-34 to M1-33 DASS-6 to DASS-3	*Glatziella* (FAD) *Kamptoclymenia* (FAD) *Parawocklumeria* (FAD) *Muessenbiaergia* (LAD)	
			...

UA	Horizons	Defining species	Remarks
...			
UA 8	M1-31 to M1-26 M4B-6 DASS-1	*Balvia globulare* (FAD) *Discoclymenia* (FAD) *Lissoclymenia* (FAD)	
UA 9	ORSTA-7a(1) ORSTB-6b(3)	*Kenseyoceras* (FAD) *Kalloclymenia* (LAD)	
UA 10	ORBK-6 to ORBK-5 ORSK-5 to ORSK-4 M1-15 M4A-18 to M4A-16 M4B-4 to M4B-1 DASM-3	*Finiclymenia* (FAD) *Postglatziella* (FAD) *Effenbergia* (LAD) *Kosmoclymenia* (LAD)	
UA 11	ORBK-4 to ORBK-1a ORSK-2 to ORSK-1 ORSTA-3(3) to ORSTA-1(1) M1-14 to M1-4 M4A-8 to M4A-4 DASM-1-b DD-93 to DD-90	*Wocklumeria* (EXC)	

Tab. 18: Description of the unitary associations obtained by Analysis G of the Devonian dataset, including the horizons, which can be assigned to the UAs. Genera can define a UA by (1) an exclusive occurrence (EXC), (2) a first appearance (FAD) or (3) a last appearance (LAD).

Reliability – 94 of 283 horizons can be clearly assigned to UAs.

Carboniferous

The calculation of the UAs was carried out with different datasets to test which size of dataset and also which modifications lead to the best result (Tab. 19).

Analysis	Dataset	Unitary Associations	Maximal Cliques	Contradictions
A	ORBV	9	9	0
B	ORBV + ORBW	11	14	11
C	ORBV + ORBW + BO + H + M2	13	19	36
D	Complete	13	20	37
E	FAD only	18	21	26
F	singletons omitted	10	13	8
G	Genus level	4	5	0

Tab 19: Overview over the analyses of the Carboniferous dataset.

Analysis A

Analysed dataset: Oberrödinghausen railway cutting section by Vöhringer (1960) (ORBV) (Fig. 20; Tab. 20).

Result in numbers: 9 unitary associations, 9 maximal cliques and 0 contradictions.

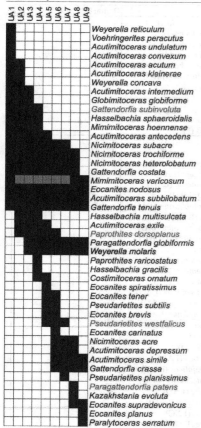

Fig. 20: Result of Analysis A of the Carboniferous dataset.

UA	Horizons	Defining species	Remarks
UA 1	ORBV-6 to ORBV-5	*Weyerella reticulum* (EXC)	The horizons ORBV-6 and
		Voehringerites peracutus (EXC)	ORBV-5 can not be
		Acutimitoceras undulatum (EXC)	distinguished, because in
		Acutimitoceras convexum (EXC)	the older horizon ORBV-6
		Acutimitoceras kleinerae (FAD)	only taxa occur, which did
		Acutimitoceras acutum (FAD)	not go extinct in this
		Weyerella concava (FAD)	horizon, but are still present
		Acutimitoceras intermedium (FAD)	in ORBV-5.
		Globimitoceras globiforme (FAD)	
		Gattendorfia subinvoluta (FAD)	...

UA	Horizons	Defining species	Remarks
...		*Hasselbachia sphaeroidalis* (FAD) *Mimimitoceras hoennense* (FAD) *Acutimitoceras antecedens* (FAD) *Nicimitoceras subacre* (FAD) *Nicimitoceras heterolobatum* (FAD) *Nicimitoceras trochiforme* (FAD) *Gattendorfia costata* (FAD) *Mimimitoceras varicosum* (FAD) *Eocanites nodosus* (FAD) *Acutimitoceras subbilobatum* (FAD) *Gattendorfia tenuis* (FAD)	
UA 2	ORBV-4	*Acutimitoceras exile* (FAD) *Hasselbachia multisulcata* (FAD) *Paprothites dorsoplanus* (FAD) *Acutimitoceras acutum* (LAD) *Acutimitoceras kleinerae* (LAD) *Weyerella concava* (LAD)	
UA 3	ORBV-3e	*Paragattendorfia globiformis* (FAD) *Weyerella molaris* (FAD) *Acutimitoceras intermedium* (LAD) *Globimitoceras globiforme* (LAD)	
UA 4	ORBV-3d	*Hasselbachia gracilis (EXC)* *Paprothites raricostatus (EXC)* *Costimitoceras ornatum* (FAD) *Gattendorfia subinvoluta* (LAD) *Hasselbachia sphaeroidalis* (LAD) *Hasselbachia multisulcata* (LAD) *Mimimitoceras hoennense* (LAD)	
UA 5	ORBV-3c	*Eocanites spiratissimus* (EXC) *Eocanites tener* (FAD) *Eocanites brevis* (FAD) *Pseudarietites subtilis* (FAD) *Pseudarietites westfalicus* (FAD) *Acutimitoceras antecedens* (LAD) *Acutimitoceras exile* (LAD) *Costimitoceras ornatum* (LAD)	
UA 6	ORBV-3b	*Eocanites carinatus* (EXC) *Nicimitoceras acre* (FAD) *Acutimitoceras depressum* (FAD) *Acutimitoceras simile* (FAD) *Gattendorfia crassa* (FAD) *Paprothites dorsoplanus* (LAD) *Eocanites tener* (LAD) *Eocanites brevis* (LAD) *Pseudarietites subtilis* (LAD)	
UA 7	ORBV-3a	*Pseudarietites planissimus* (EXC) *Nicimitoceras subacre* (LAD) *Pseudarietites westfalicus* (LAD)	
UA 8	ORBV-2	*Parawocklumeria patens* (EXC) *Kazakhstania evoluta* (EXC) *Eocanites supradevonicus* (FAD) *Nicimitoceras trochiforme* (LAD) *Nicimitoceras heterolobatum* (LAD) *Gattendorfia costata* (LAD)	...

UA	Horizons	Defining species	Remarks
...		*Nicimitoceras acre* (LAD)	
		Acutimitoceras depressum (LAD)	
UA 9	ORBV-1	*Eocanites planus* (EXC)	
		Paralytoceras serratum (EXC)	

Tab 20: Description of the unitary associations obtained by Analysis A of the Carboniferous dataset, including the horizons, which can be assigned to the UAs. Species can define a UA by (1) an exclusive occurrence (EXC), (2) a first appearance (FAD) or (3) a last appearance (LAD).

Reliability – All ten horizons can be clearly assigned to UAs.

Analysis B

Analysed dataset: Oberrödinghausen railway cutting sections studied by Vöhringer (1960) (ORBV) and Weyer (ORBW), of which the latter is a section adjacent to ORBV (Fig. 21; Tab. 21).

Result in numbers: 11 unitary associations, 14 maximal cliques and 11 contradictions.

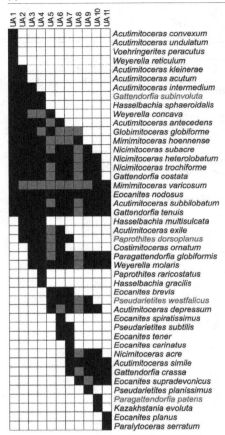

Fig. 21: Result of Analysis B of the Carboniferous dataset.

UA	Horizons	Defining species	Remarks
UA 1	ORBW-5c to ORBW-5b ORBV-6 to ORBV-5	*Weyerella reticulum* (EXC) *Voehringerites peracutus* (EXC) *Acutimitoceras undulatum*(EXC) *Acutimitoceras convexum* (EXC) *Acutimitoceras kleinerae* (FAD) *Acutimitoceras acutum* (FAD) *Acutimitoceras intermedium* (FAD) *Gattendorfia subinvoluta*(FAD) *Hasselbachia sphaeroidalis* (FAD) *Weyerella concava* (FAD)	...

UA	Horizons	Defining species	Remarks
...		*Acutimitoceras antecedens* (FAD) *Globimitoceras globiforme* (FAD) *Mimimitoceras hoennense* (FAD) *Nicimitoceras subacre* (FAD) *Nicimitoceras heterolobatum* (FAD) *Nicimitoceras trochiforme* (FAD) *Gattendorfia costata* (FAD) *Mimimitoceras varicosum* (FAD) *Eocanites nodosus* (FAD) *Acutimitoceras subbilobatum* (FAD) *Gattendorfia tenuis* (FAD)	
UA 2	ORBV-4	*Acutimitoceras exile* (FAD) *Hasselbachia multisulcata* (FAD) *Paprothites dorsoplanus (FAD)* *Acutimitoceras acutum* (LAD) *Acutimitoceras kleinerae* (LAD)	
UA 3	ORBW-3e ORBV-3e	*Costimitoceras ornatum* (FAD) *Paragattendorfia globiformis* (FAD) *Weyerella molaris* (FAD) *Acutimitoceras acutum* (LAD)	
UA 4	ORBW-3d	*Hasselbachia gracilis (EXC)* *Paprothites raricostatus (EXC)* *Gattendorfia subinvoluta* (LAD) *Hasselbachia sphaeroidalis* (LAD)	
UA 5	ORBW-3c2 to ORBW-3a	*Acutimitoceras depressum* (FAD) *Eocanites brevis* (FAD) *Pseudarietites westfalicus* (FAD) *Weyerella concava* (LAD)	The LAD of *Weyerella concava* is reported from Weyer in bed ORBW-3a but from Vöhringer in bed ORBV-5, which makes this UA only partially reliable.
UA 6	ORBV-3c	*Eocanites spiratissimus (EXC)* *Eocanites tener* (FAD) *Pseudarietites subtilis* (FAD) *Acutimitoceras antecedens* (LAD) *Acutimitoceras exile* (LAD) *Costimitoceras ornatum* (LAD)	
UA 7	ORBV-3b	*Eocanites carinatus* (EXC) *Eocanites brevis* (LAD) *Eocanites tener* (LAD) *Pseudarietites subtilis* (LAD) *Paprothites dorsoplanus* (LAD)	
UA 8	ORBW-2b	*Eocanites supradevonicus* (FAD) *Mimimitoceras hoennense* (LAD)	The LAD of *Mimimitoceras hoennense* is in ORBW-2a and ORBV-3d. So UA 8 can be regarded as only partially reliable.
UA 9	ORBV-3a	*Pseudarietites planissimus* (EXC) *Nicimitoceras subacre* (LAD) *Pseudarietites westfalicus* (LAD)	
			...

UA	Horizons	Defining species	Remarks
...			
UA 10	ORBV-2	*Paragattendorfia patens* (EXC) *Kazakhstania evoluta* (EXC) *Eocanites supradevonicus* (FAD) *Nicimitoceras trochiforme* (LAD) *Nicimitoceras heterolobatum* (LAD) *Gattendorfia costata* (LAD) *Nicimitoceras acre* (LAD) *Acutimitoceras depressum* (LAD)	
UA 11	ORBV-1	*Eocanites planus* (EXC) *Paralytoceras serratum* (EXC)	Weyer did not sample bed 1

Tab 21: Description of the unitary associations obtained by Analysis B of the Carboniferous dataset, including the horizons, which can be assigned to the UAs. Species can define a UA by (1) an exclusive occurrence (EXC), (2) a first appearance (FAD) or (3) a last appearance (LAD).

Reliability – 18 of the 25 horizons can be clearly assigned to UAs.

Comparison of Analysis A and Analysis B – The species mostly show the same succession. It is noticeable that the taxa *Weyerella concava* and *Mimimitoceras hoennense* have their LAD later in Analysis B because of discontinuities. The occurrence of *Costimitoceras ornatum* is prolonged from 2 to 4 UAs in Analysis B because it occurs in the horizons ORBW-3e, ORBV-3c and ORBV-3d. *Eocanites supradevonicus* has its calculated FAD in UA 8, because it occurs in ORBW-2b and ORBV-2. *Acutimitoceras depressum* exhibits an elongated range, which can only be explained by co-occurences. So the addition of ORBW added more uncertainties than improvement to the stratigraphical resolution. Additionally it can be said, that Weyer's subdivision can not be confirmed via the UA method.

Analysis C

Analysed dataset: Oberrödinghausen railway cutting sections studied by Vöhringer (1960) (ORBV) and Weyer (ORBW) + Becke-Oese (BO), Hasselbach (H) and Müssenberg (M2) (Fig. 22; Tab. 22).

Result in numbers: 13 unitary associations, 19 maximal cliques and 36 contradictions.

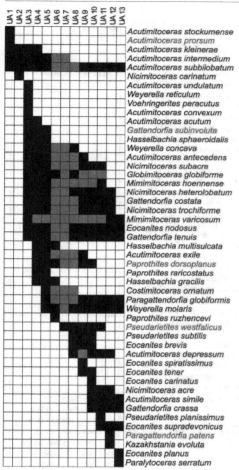

Fig. 22: Result of Analysis C of the Carboniferous dataset.

UA	Horizons	Defining species	Remarks
UA 1	M2-3A	Acutimitoceras procedens (EXC)	
		Acutimitoceras stockumense (EXC)	
		Acutimitoceras intermedium (FAD)	
		Acutimitoceras kleinerae (FAD)	
		Acutimitoceras subbilobatum (FAD)	
			...

UA	Horizons	Defining species	Remarks
...			
UA 2	H-83 M2-3B	*Nicimitoceras carinatum* (EXC)	
UA 3	H-72 to H-59 ORBW-5c to ORBW-5b ORBV-6 to ORBV-5	*Weyerella reticulum* (EXC) *Voehringerites peracutus* (EXC) *Acutimitoceras undulatum*(EXC) *Acutimitoceras convexum* (EXC) *Acutimitoceras acutum* (FAD) *Gattendorfia subinvoluta*(FAD) *Hasselbachia sphaeroidalis* (FAD) *Weyerella concava* (FAD) *Acutimitoceras antecedens* (FAD) *Nicimitoceras subacre* (FAD) *Globimitoceras globiforme* (FAD) *Mimimitoceras hoennense* (FAD) *Nicimitoceras heterolobatum* (FAD) *Nicimitoceras trochiforme* (FAD) *Gattendorfia costata* (FAD) *Mimimitoceras varicosum* (FAD) *Eocanites nodosus* (FAD) *Eocanites nodosus* (FAD) *Gattendorfia tenuis* (FAD)	Can be seen as equal to UA 1 in the reference section.
UA 4	ORBV-4	*Acutimitoceras exile* (FAD) *Hasselbachia multisulcata* (FAD) *Paprothites dorsoplanus (FAD)* *Acutimitoceras acutum* (LAD)	
UA 5	ORBW-3d1 ORBV-3e to ORBV-3d	*Paprothites raricostatus* (EXC) *Costimitoceras ornatum* (FAD) *Hasselbachia gracilis* (FAD) *Paragattendorfia globiformis* (FAD) *Weyerella molaris* (FAD) *Gattendorfia subinvoluta* (LAD) *Hasselbachia sphaeroidalis* (LAD)	
UA 6	H-57	*Paprothites ruzhencevi* (EXC)	Contains many discontinuities and hence adds many uncertainties.
UA 7	H-49	*Pseudarietites westfalicus* (FAD)	Contains many discontinuities and hence adds many uncertainties. *Pseudarietites westfalicus* has 9 occurrences in total and does not show any discontinuities. Hence, although the rest of UA 7 is only defined by discontinuities, it seems rather reliable. Equals UA 5 in Analysis A.
			...

UA	Horizons	Defining species	Remarks
...			
UA 8	BO-28 ORBW-3c1 to ORBW-3a	*Acutimitoceras depressum* (FAD) *Pseudarietites subtilis* (FAD) *Eocanites brevis* (FAD) *Weyerella concava (LAD)*	Equals UA 5 in Analysis A.
UA 9	ORBV-3c	*Eocanites spiratissimus* (EXC) *Eocanites tener* (FAD) *Acutimitoceras exile* (LAD) *Costimitoceras ornatum* (LAD)	Equals UA 5 in Analysis A. In total UA 5 of the reference section is subdivided into three UAs in this analysis.
UA 10	ORBV-3b	*Eocanites carinatus* (EXC) *Nicimitoceras acre* (FAD) *Acutimitoceras simile* (FAD) *Gattendorfia crassa* (FAD) *Paprothites dorsoplanus* (LAD) *Eocanites tener* (LAD) *Eocanites brevis* (LAD) *Pseudarietites subtilis* (LAD)	UA 6 in the reference section.
UA 11	ORBW-2b ORBV-3a	*Pseudarietites planissimus* (EXC) *Eocanites supradevonicus* (FAD) *Pseudarietites westfalicus* (LAD)	UA 7 in the reference section.
UA 12	BO-30 ORBV-2	*Paragattendorfia patens* (EXC) *Kazakhstania evoluta* (EXC) *Acutimitoceras depressum* (LAD) *Nicimitoceras acre* (LAD) *Nicimitoceras trochiforme* (LAD) *Nicimitoceras heterolobatum* (LAD) *Gattendorfia costata* (LAD)	UA 8 in the reference section.
UA 13	ORBV-1	*Eocanites planus* (EXC) *Paralytoceras serratum* (EXC)	UA 9 in the reference section.

Tab 22: Description of the unitary associations obtained by Analysis C of the Carboniferous dataset, including the horizons, which can be assigned to the UAs. Species can define a UA by (1) an exclusive occurrence (EXC), (2) a first appearance (FAD) or (3) a last appearance (LAD).

Reliability – 27 of the 50 horizons can be clearly assigned to UAs.

Comparison of Analysis A and Analysis C – The most noticeable differences between the results of Analysis A and Analysis C are: (1) UA 3 and 4 of Analysis A form UA 5 in Analysis C and (2) the FAD of *Acutimitoceras depressum* is earlier in Analysis C.

Analysis D

Analysed dataset: The complete corrected Carboniferous dataset (Fig. 23; Tab. 23).

Result in numbers: 13 unitary associations, 20 maximal cliques and 37 contradictions.

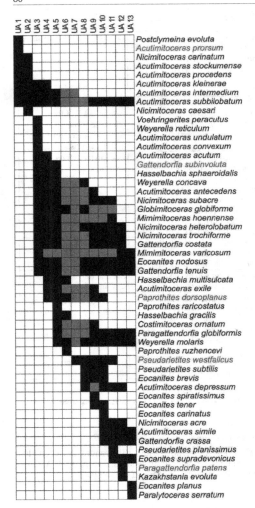

Fig. 23: Result of Analysis D of the Carboniferous dataset (from Klein and Korn 2015).

UA	Horizons	Defining species	Remarks
UA 1	M2-3A	*Acutimitoceras procedens* (EXC) *Postclymeina evoluta* (EXC) *Nicimitoceras carinatum* (FAD) *Acutimitoceras stockumense* (FAD) *Acutimitoceras prorsum* (FAD) *Acutimitoceras intermedium* (FAD) *Acutimitoceras kleinerae* (FAD) *Acutimitoceras subbilobatum* (FAD)	It contains three new taxa compared to Analysis C (*Postclymenia evoluta, Nicimitoceras carinatum, Acutimitoceras prorsum*).
UA 2	S-K	*Nicimitoceras caesari* (EXC) *Nicimitoceras carinatum* (LAD) *Acutimitoceras stockumense* (LAD) *Acutimitoceras prorsum* (LAD)	
The taxa in UA 3 to UA 13 remain unchanged with respect to Analysis C. The horizon DK-4a adds to UA 3. For the UAs 4 to 13 the content of the horizons remains unchanged with respect to Analysis C.			

Tab 23: Description of the unitary associations obtained by Analysis D of the Carboniferous dataset, including the horizons, which can be assigned to the UAs. Species can define a UA by (1) an exclusive occurrence (EXC), (2) a first appearance (FAD) or (3) a last appearance (LAD).

Reliability – 29 of 59 horizons can be clearly assigned to UAs.

Stratigraphical order of the sections – Müssenberg (UA 1) is the lowermost section in the complete succession, followed by Drewer (UA 1, UA 3) and Stockum (UA 2). Hasselbach, ORBW and ORBV all start with UA 3, Hasselbach reaching to UA 7, ORBW to UA 11 and ORBV is the uppermost section of the complete succession ranging to UA 13. Becke-Oese is only represented in two UAs (UA 8 and UA 12).

Comparison of Analysis A and Analysis D – The same ammonoid associations occur in Analysis A and Analysis D. The most noticeable differences are: (1) UA 3 and UA 4 of Analysis A form UA 5 in Analysis D and (2) the FAD of *Acutimitoceras depressum* is earlier in Analysis D.

Analysis E

Analysed dataset: The complete corrected Carboniferous dataset with first occurrences only (Fig. 24; Tab. 24).

Result in numbers: 18 unitary associations, 21 maximal cliques and 26 contradictions. Additionally 4 Cliques in Cycles and 9 residual virtual edges appear. Those cycles are also called forbidden sub-graphs. In this case S3 circuits, that is "A co-occurring with B, C above A and C below B" (Guex 1991; Hammer 2012).

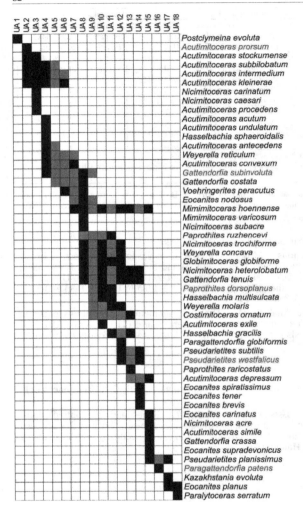

Fig. 24: Result of Analysis E of the Carboniferous dataset.

UA	Horizons	Defining species	Remarks
UA 1	DK-99	*Postclymeina evoluta* (FAD)	
			...

UA	Horizons	Defining species	Remarks
...			
UA 2	M2-3A	*Acutimitoceras procedens* (FAD) *Acutimitoceras stockumense* (FAD) *Acutimitoceras intermedium* (FAD) *Acutimitoceras kleinerae* (FAD) *Acutimitoceras subbilobatum* (FAD)	
UA 3	H-83 M2-3B SK DK-100 to DK-1	*Acutimitoceras prorsum* (FAD) *Nicimitoceras caesari* (FAD) *Nicimitoceras carinatum* (FAD)	It is not possible to clearly dissolve the Stockum limestone.
UA 4	DK-5 to DK-4a ORBW-6b ORBV-6	*Acutimitoceras acutum* (FAD) *Acutimitoceras undulatum* (FAD) *Hasselbachia sphaeroidalis* (FAD) *Acutimitoceras antecedens* (FAD) *Acutimitoceras convexum* (FAD) *Gattendorfia subinvoluta* (FAD) *Weyerella reticulum* (FAD)	
UA 5	Is entirely composed of discontinuities and hence without a source for stratigraphical information.		
UA 6	ORBW-5c	*Voehringerites peracutus* (FAD) *Acutimitoceras kleinerae* (LAD)	Mainly composed of discontinuities. Contains only two taxa.
UA 7	H-59	*Mimimitoceras hoennense* (FAD)	Mainly composed of discontinuities. Contains only two taxa.
UA 8	BO-11 ORBV-5	*Mimimitoceras varicosum* (FAD) *Nicimitoceras subacre* (FAD) *Gattendorfia tenuis* (FAD) *Globimitoceras globiforme* (FAD) *Nicimitoceras heterolobatum* (FAD) *Nicimitoceras trochiforme* (FAD) *Weyerella concava* (FAD)	
UA 9	Is entirely composed of discontinuities and hence without a source for stratigraphical information.	UA 9	Is entirely composed of discontinuities and hence without a source for stratigraphical information.
UA 10	ORBV-4	*Acutimitoceras exile* (FAD)	It can be seen as equivalent to UA 4 in Analysis D.
UA 11	H-57	*Hasselbachia gracilis* (FAD)	It can be seen as equivalent to UA 6 in Analysis D.
UA 12	BO-28 ORBV-3e	*Paragattendorfia globiformis* (FAD) *Pseudarietites subtilis* (FAD) *Pseudarietites westfalicus* (FAD)	Interestingly, the two horizons belong to UA 5 (ORBV-3e) and UA 8 (BO-28) in Analysis D.
UA 13	ORBV-3d	*Paprothites raricostatus* (FAD)	ORBV-3d also belongs to UA 5 in Analysis D.
UA 14	ORBW-3c1 ORBV-3c .	*Eocanites brevis* (FAD) *Eocanites spiratissimus* (FAD) *Eocanites tener* (FAD)	In Analysis D these horizons belong to different UAs (ORBV-3c to UA 9, ORBW-3c1 to UA 8).
			...

UA	Horizons	Defining species	Remarks
...			
UA 15	ORBW-2b ORBV-3b	*Eocanites carinatus* (FAD) *Nicimitoceras acre* (FAD) *Acutimitoceras simile* (FAD) *Gattendorfia crassa* (FAD) *Eocanites supradevonicus* (FAD)	In Analysis D these horizons belong to different UAs (ORBV-3b to UA 10, ORBW-2b to UA 11).
UA 16	ORBV-3a	*Pseudarietites planissimus* (FAD)	ORBV-3a belonged to UA 11 in Analysis D.
UA 17	BO-30 ORBV-2	*Paragattendorfia patens* (FAD) *Kazakhstania evoluta* (FAD)	It is equivalent to UA 12 in Analysis D.
UA 18	ORBV-1	*Eocanites planus* (FAD) *Paralytoceras serratum* (FAD)	

Tab 24: Description of the unitary associations obtained by Analysis E of the Carboniferous dataset, including the horizons, which can be assigned to the UAs. Species can define a UA by (1) an exclusive occurrence (EXC) or (2) a first appearance (FAD).

Reliability – 22 of 59 horizons can be clearly assigned to UAs.

Comparison of Analysis D and Analysis E – It can be clearly seen that, using the two different approaches, the horizons and species are assigned to different UAs. One reason for the additional UAs compared to Analysis D is the fact, that there is only one criterion, the first occurrence, used to distinguish the associations. For example for the ORBV section the horizons ORBV-5 and ORBV-6 can not be distinguished using FADs and LADs, because in the horizon ORBV-6 only taxa occur, which are still present in ORBV-5. Using FADs only, the horizons can be distinguished, because new taxa occur in ORBV-5. Two Associations (UA 5 and UA 9), which contain only discontinuities occur, because of the missing criterion. Additionally it is more difficult to correlate the sections, because they do not fit together perfectly.

Analysis F

Analysed dataset: The complete corrected Carboniferous dataset with singletons omitted (Fig. 25; Tab. 25).

Result in numbers: 10 unitary associations, 13 maximal cliques and 8 contradictions.

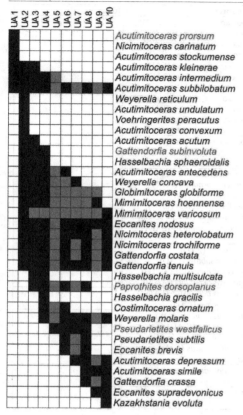

Fig. 25: Result of Analysis F of the Carboniferous dataset.

UA	Horizons	Defining species	Remarks
UA 1	H-83	*Acutimitoceras prorsum* (EXC)	
	M2-3B to M2-3A	*Acutimitoceras stockumense* (EXC)	
	S-K	*Nicimitoceras carinatum* (EXC)	
	DK-100 to DK-1	*Acutimitoceras intermedium* (FAD)	
		Acutimitoceras kleinerae (FAD)	
		Acutimitoceras subbilobatum (FAD)	
UA 2	H-72 to	*Weyerella reticulum* (EXC)	
	H-59	*Voehringerites peracutus* (EXC)	
	ORBW-5c to ORBW-5b	*Acutimitoceras undulatum* (EXC)	
	ORBV-6 to ORBV-5	*Acutimitoceras convexum* (EXC)	
	DK-4a	*Acutimitoceras acutum* (FAD)	
		Gattendorfia subinvoluta (FAD)	...

UA	Horizons	Defining species	Remarks
...		*Hasselbachia sphaeroidalis* (FAD)	
		Weyerella concava (FAD)	
		Acutimitoceras antecedens (FAD)	
		Globimitoceras globiforme (FAD)	
		Mimimitoceras hoennense (FAD)	
		Mimimitoceras varicosum (FAD)	
		Eocanites nodosus (FAD)	
		Nicimitoceras heterolobatum (FAD)	
		Nicimitoceras trochiforme (FAD)	
		Gattendorfia costata (FAD)	
		Gattendorfia tenuis (FAD)	
UA 3	ORBV-4	*Hasselbachia multisulcata* (FAD)	
		Paprothites dorsoplanus (FAD)	
		Acutimitoceras acutum (LAD)	
UA 4	H-57	*Hasselbachia gracilis* (EXC)	
	ORBW-3e to ORBW-3d1	*Costimitoceras ornatum* (FAD)	
	ORBV-3e to ORBV-3d	*Weyerella molaris* (FAD)	
		Gattendorfia subinvoluta (LAD)	
		Hasselbachia sphaeroidalis (LAD)	
UA 5	H-49	*Pseudarietites westfalicus* (FAD)	Mainly composed of discontinuities.
UA 6	ORBV-3c	*Eocanites brevis* (FAD)	
		Pseudarietites subtilis (FAD)	
		Acutimitoceras antecedens (LAD)	
		Costimitoceras ornatum (LAD)	
UA 7	ORBW-3a	*Acutimitoceras depressum* (FAD)	
		Weyerella concava (LAD)	
UA 8	ORBV-3b to ORBV-3a	*Acutimitoceras simile* (FAD)	
		Gattendorfia crassa (FAD)	
		Paprothites dorsoplanus (LAD)	
		Pseudarietites subtilis (LAD)	
		Eocanites brevis (LAD)	
UA 9	ORBW-2b	*Eocanites supradevonicus* (FAD)	
		Mimimitoceras hoennense (LAD)	
UA 10	BO-30	*Kazakhstania evoluta* (EXC)	
	ORBV-2		

Tab 25: Description of the unitary associations obtained by Analysis F of the Carboniferous dataset, including the horizons, which can be assigned to the UAs. Species can define a UA by (1) an exclusive occurrence (EXC), (2) a first appearance (FAD) or (3) a last appearance (LAD).

Reliability – Only 29 horizons of the 59 horizons can be clearly assigned to UAs. The first thing that becomes apparent, is that 16 of the 52 species are only represented in one section, containing all species, which were only present in SK. For that reason, the dataset omitting singletons will not be used for the further analysis in order not to lose the information about 16 species. Also, together with the 16 species, there is some relevant stratigraphical information which is lost. In fact, the succession of the UAs and their taxa did not change noticeably, but due to the many missing species, the resolution is coarser.

Comparison of Analysis D and Analysis F – The structure of the UAs in Analysis F is the same as in Analysis D. Only the resolution is lower due to the 16 missing singletons.

Analysis G

Analysed dataset: The complete corrected Carboniferous on genus level. 15 genera are included, but 5 genera are only defined by one species (Fig. 26).

Result in numbers: 4 unitary associations, 5 maximal cliques and 0 contradictions.

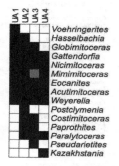

Fig. 26: Result of Analysis G of the Carboniferous dataset.

Description of the unitary associations: Two of these unitary associations are defined by *Voehringerites peracutus* and *Kazakhstania evoluta*. When those single species are removed, only two unitary associations are left, which makes it obvious, that the resolution is too coarse on genus level.

5.2 Constrained Optimization

The most important output file of the CONOP analysis is the composite section (compst.dat), which shows the FADs and LADs of the complete succession for all taxa. In order to make the compst.dat file more readable, I visualized the ranges of the individual taxa assigning the taxon names to the event numbers and sorting the data ascending after their CONOP FAD and afterwards after their CONOP LAD. Unfortunately the time needed for the compilation of the input files as well as the calculations with CONOP9 is quite substantial, so that it is difficult to try different approaches.

Devonian
Stratigraphical arrangement

87 CONOP units, which include a total of 64 species, were calculated for the Devonian dataset (Fig. 27). 20 of these species do not occur in the reference dataset Müssenberg 1 (M1) with 44 species. In general the result of the CONOP analysis coincides with the reference section. Only those taxa, which have their FAD in the EHs 15 to 23 in the reference section, are difficult to separate by CONOP. Additionally, the FADs of *Effenbergia lens* and *Mimimitoceras trizonatum* are postponed in the results of the CONOP analysis.

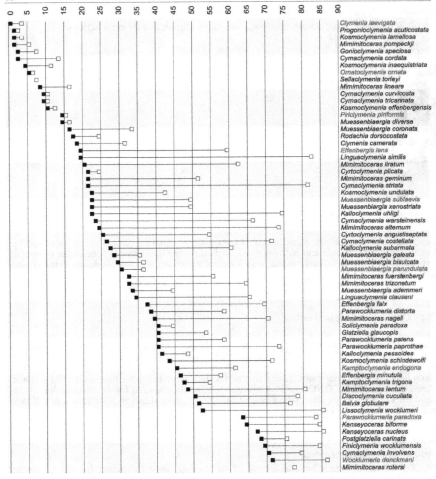

Fig. 27: Result of the CONOP analysis of the Devonian dataset (from Klein and Korn 2015).

Carboniferous

Stratigraphical arrangement

35 CONOP units were calculated for a total of 52 species of the Carboniferous dataset (Fig. 28). Seven of these species do not appear in the reference section Oberrödinghausen railway cutting by Vöhringer (1960) (ORBV). There is a good agreement between the reference section and the result of the CONOP analysis. Only for *Paprothites dorsoplanus*,

Gattendorfia costata and *Voehringerites peracutus* are the CONOP FADs earlier than the FADs in the reference section.

Fig. 28: Result of the CONOP analysis of the Carboniferous dataset (from Klein and Korn 2015).

5.3   Ranking and Scaling

The implementation of RASC in PAST offers three options when samples are transcribed to events, of which two were used: Using FADs and LADs and using only FADs (Hammer 2012). The ranges of the taxa were visualized by sorting them ascending after their RASC FAD and afterwards after their RASC LAD.

Devonian
Stratigraphical arrangement with FADs and LADs

The 65 species of the complete Devonian dataset form 130 RASC units (Tab. 26 ; Fig. 29). The FADs and LADs of the RASC analysis are given with their error bar lengths. In general the error bars seldom exceed the length of 5 units and never the length of 10 units. Species that are especially reliable for stratigraphic ordering possesses error bar lengths under 5 units and ranges under 10 units.

Species	E	FAD	E		E	LAD	E	Remarks
Clymenia laevigata	0	1	0		0	2	4	Reliable for stratigraphic ordering
Kosmoclymenia lamellosa	1	3	6		2	4	5	Maximum error bar length: six units.
Progonioclymenia acuticostata	3	5	4		4	6	3	Reliable for stratigraphic ordering
Cymaclymenia cordata	4	7	2		0	25	0	
Mimimitoceras pompeckji	7	8	2		8	9	1	Maximum error bar length: eight units.
Gonioclymenia speciosa	2	10	1		0	14	0	
Kosmoclymenia inaequistriata	1	11	0		1	20	3	
Ornatoclymenia ornata	0	12	0		0	13	0	Reliable for stratigraphic ordering
Mimimitoceras lineare	0	15	2		0	18	0	Reliable for stratigraphic ordering
Sellaclymenia torleyi	4	16	1		5	17	0	Maximum error bar length: five units.
Cyrtoclymenia tetragona	0	19	4		2	21	2	Reliable for stratigraphic ordering
Cymaclymenia curvicosta	3	22	1		4	23	0	Reliable for stratigraphic ordering
Kosmoclymenia effenbergensis	0	24	0		4	29	0	Reliable for stratigraphic ordering
Muessenbiaergia diversa	1	27	4		0	26	5	Calculated FAD before calculated LAD.
Muessenbiaergia coronata	2	28	3		0	44	0	
Piriclymenia piriformis	0	30	0		0	31	0	Reliable for stratigraphic ordering
Linguaclymenia similis	0	32	3		0	124	0	
Muessenbiargia xenostriata	1	33	0		0	96	0	
Rodachia dorsocostata	1	35	1		0	34	2	Calculated FAD before calculated LAD.
Muessenbiaergia sublaevis	3	36	1		0	72	1	
Cymaclymenia camerata	1	37	0		1	59	0	
Mimimitoceras liratum	0	38	0		0	97	1	
Cyrtoclymenia plicata	0	39	1		0	41	0	Reliable for stratigraphic ordering
Cymaclymenia striata	1	40	0		1	111	0	
Kosmoclymenia undulata	0	42	0		0	65	0	
Cymaclymenia warsteinensis	0	43	0		3	98	0	
Cyrtoclymenia angustiseptata	0	45	0		2	85	0	
Cymaclymenia costellata	0	46	1		0	64	0	
Kalloclymenia uhligi	5	47	0		0	129	1	
Muessenbiaergia galeata	0	48	1		1	51	1	Reliable for stratigraphic ordering
Mimimitoceras geminum	2	49	1		0	80	0	
Muessenbiaergia bisulcata	1	50	3		0	54	0	
Mimimitoceras alternum	2	52	2		8	58	3	Maximum error bar length: eight units.
Muessenbiaergia parundulata	1	53	0		2	57	1	Reliable for stratigraphic ordering
Muessenbiaergia ademmeri	0	55	0		1	68	0	
Linguaclymenia clauseni	0	56	5		0	95	0	
Effenbergia lens	0	60	0		0	90	0	
Mimimitoceras fuerstenbergi	0	61	0		0	86	0	
Kalloclymenia subarmata	0	62	0		1	88	0	
Mimimitoceras trizonatum	0	63	0		2	70	0	
Soliclymenia paradoxa	0	66	0		0	67	1	Reliable for stratigraphic ordering
Glatziella glaucopis	0	69	1		0	91	0	
Parawocklumeria patens	0	71	0		0	89	0	
Kalloclymenia pessoides	1	73	2		0	83	2	Reliable for stratigraphic ordering
Effenbergia falx	1	74	0		1	101	1	
								...

Species	E	FAD	E		E	LAD	E	Remarks
...								
Parawocklumeria distorta	0	75	0		0	92	2	
Kosmoclymenia schindewolfi	0	76	1		0	104	0	
Mimimitoceras nageli	1	77	0		0	87	1	
Kamptoclymenia endogona	0	78	0		0	94	0	
Parawocklumeria paprothae	0	79	0		3	102	2	
Kamptoclymenia trigona	1	82	0		0	81	1	Calculated FAD before calculated LAD.
Effenbergia minutula	3	84	2		4	93	0	Reliable for stratigraphic ordering
Mimimitoceras lentum	0	99	0		0	123	0	
Discoclymenia cucullata	0	100	2		6	107	2	Maximum error bar length: six units.
Lissoclymenia wocklumeri	1	103	0		0	128	0	
Kenseyoceras biforme	0	105	2		0	118	0	
Parawocklumeria paradoxa	1	106	2		0	125	1	
Balvia globulare	2	108	0		0	114	0	Reliable for stratigraphic ordering
Postglatziella carinata	0	109	1		0	116	0	
Finiclymenia wocklumensis	2	110	1		1	126	0	
Kenseyoceras nucleus	2	113	0		0	127	0	
Cymaclymenia involvens	0	115	0		1	122	0	
Wocklumeria denckmanni	0	117	0		1	130	0	
Mimimitoceras rotersi	0	119	2		1	120	1	Reliable for stratigraphic ordering

Tab. 26: FADs and LADs with error bars of the species obtained by the RASC analysis of the Devonian dataset [E=Error bar length].

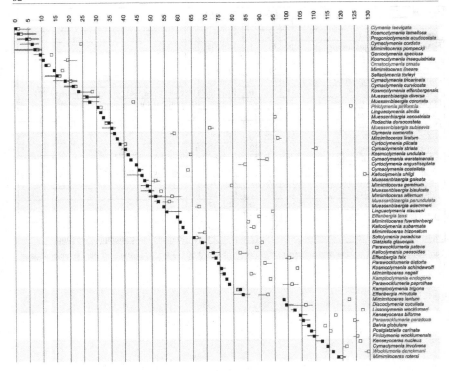

Fig. 29: Result of the RASC analysis including FADs and LADs with error bars of the Devonian dataset (from Klein and Korn 2015).

Comparison with reference section

In general the ammonoid succession obtained by the RASC analysis shows a good congruency with the reference section Müssenberg 1 (M1). The taxa, which have their FAD in the EHs 15 to 18 in the reference section seem hard to resolve via RASC: The RASC FAD of *Mimimitoceras liratum* and *Cymaclymenia striata* is earlier. The RASC FAD of *Kalloclymenia subarmata*, which has its FAD in EH 19 of the reference section, is postponed. In the result of the RASC analysis, *Kalloclymenia subarmata* occurs together with *Mimimitoceras fuerstenbergi* (error bar length 0), which occurs much later (EH 30) in the reference dataset. In addition, the taxa, which have their FAD in the EHs 26 to 28 in the reference section, seem hard to resolve by the RASC method. The taxa, which have their FAD in the EHs 31 to 37 in the reference section, are hard to resolve by the RASC analysis. *Balvia globulare* with a range of 6 units and an error bar length of up to two units, seems to be rather reliable. However, the FAD of *Balvia globulare* is strongly postponed in the result of the RASC analysis.

Stratigraphical arrangement with only FADs

When only FADs are used for the analysis, the result is a straight line, on which the 65 taxa are positioned (Fig. 30). The error bar never exceeds five units in this analysis. The two analyses show only minor deviations from each other. The taxa, which can be reliably dated in the analysis with FADs and LADs, show a similar arrangement in the analysis with only FADs. In general the ammonoid succession shows a good congruency with the reference section again, the deviations are comparable to those of the former analysis.

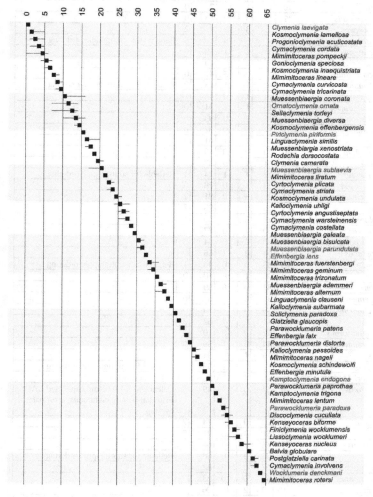

Fig. 30: Result of the RASC analysis including only FADs with error bars of the Devonian dataset.

Carboniferous

Stratigraphical arrangement with FADs and LADs

When the FADs and LADs of the complete Carboniferous dataset are analysed with RASC, 104 units arise for the 52 taxa (Tab. 27; Fig. 31). Species that are especially reliable for stratigraphic ordering possess ranges under 10 units and error bar lengths under 5 units.

Species	E	FAD	E		E	LAD	E	Remarks
Postclymeina evoluta	0	1	1		2	7	0	
Hasselbachia sphaeroidalis	1	2	5		0	43	0	
Acutimitoceras procedens	1	3	3		2	4	2	
Acutimitoceras prorsum	4	5	6		5	6	5	Reliable for stratigraphic ordering
Acutimitoceras kleinerae	0	8	2		1	19	0	
Acutimitoceras stockumense	8	9	2		9	10	1	
Acutimitoceras intermedium	2	11	0		5	47	0	
Nicimitoceras carinatum	0	12	1		1	13	0	Reliable for stratigraphic ordering
Acutimitoceras antecedens	0	14	1		0	45	14	
Acutimitoceras undulatum	1	15	1		7	31	0	
Acutimitoceras subbilobatum	1	16	2		0	93	0	
Acutimitoceras acutum	1	17	0		0	32	0	
Gattendorfia costata	0	18	1		1	81	0	
Gattendorfia subinvoluta	0	20	0		0	48	1	
Weyerella reticulum	0	21	1		0	27	0	
Voehringerites peracutus	1	22	1		1	24	1	Reliable for stratigraphic ordering
Acutimitoceras convexum	1	23	1		0	28	0	
Mimimitoceras varicosum	1	25	3		0	91	2	
Eocanites nodosus	1	26	0		1	95	7	
Globimitoceras globiforme	0	29	5		1	50	1	
Nicimitoceras subacre	6	30	1		0	77	3	
Hasselbachia multisulcata	0	33	0		5	54	3	
Paprothites dorsoplanus	0	34	0		1	66	0	
Weyerella concava	0	35	0		0	40	1	Reliable for stratigraphic ordering
Gattendorfia tenuis	0	36	1		6	104	0	
Nicimitoceras heterolobatum	1	37	0		6	89	1	
Mimimitoceras hoennense	0	38	0		2	51	0	
Nicimitoceras trochiforme	0	39	0		7	90	0	
Acutimitoceras exile	1	41	0		4	62	3	
Paragattendorfia globiformis	0	42	0		9	102	2	
Costimitoceras ornatum	0	44	0		1	59	5	
Weyerella molaris	4	46	1		0	94	1	
Hasselbachia gracilis	0	52	5		1	53	4	Reliable for stratigraphic ordering
Nicimitoceras caesari	54	55	49		55	56	48	Longest error bar
Paprothites raricostatus	6	57	0		1	49	8	Calculated FAD before calculated LAD. Error bar length: eight units, error bars overlap.
Pseudarietites subtilis	0	58	9		6	74	2	
Eocanites tener	3	61	4		3	71	5	
Eocanites spiratissimus	5	63	2		6	64	1	
Eocanites brevis	5	65	2		0	68	0	Reliable for stratigraphic ordering
Pseudarietites westfalicus	0	67	0		0	82	0	
Acutimitoceras depressum	0	69	0		0	83	7	
Acutimitoceras simile	0	70	2		1	96	6	
								...

Species	E	FAD	E		E	LAD	E	Remarks
...								
Nicimitoceras acre	4	72	4		1	84	6	
Gattendorfia crassa	2	73	3		5	103	1	
Eocanites carinatus	7	75	1		8	76	0	
Pseudarietites planissimus	1	78	2		2	79	1	Reliable for stratigraphic ordering
Paprothites ruzhencevi	28	80	1		8	60	5	Calculated FAD before calculated LAD. Error bar length: eight units, error bars overlap
Paragattendorfia patens	2	85	5		3	86	4	Reliable for stratigraphic ordering
Kazakhstania evoluta	4	87	4		5	88	3	Reliable for stratigraphic ordering
Eocanites supradevonicus	2	92	0		2	97	5	Reliable for stratigraphic ordering
Eocanites planus	6	99	5		5	98	6	
Paralytoceras serratum	7	100	4		8	101	3	

Tab. 27: FADs and LADs with error bars of the species obtained by the RASC analysis of the Carboniferous dataset [E=Error bar length].

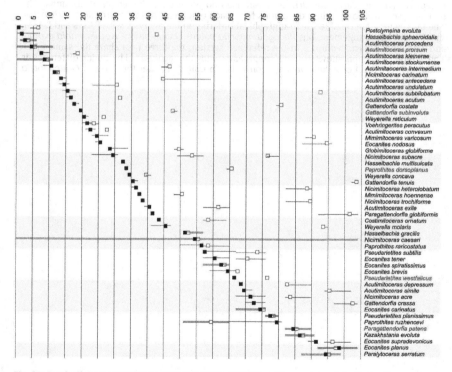

Fig. 31: Result of the RASC analysis including FADs and LADs with error bars of the Carboniferous dataset (from Klein and Korn 2015).

Comparison with reference section

In total the ammonoid succession shows a good agreement with the reference section Oberrödinghausen railway cutting by Vöhringer (1960) (ORBV). The species, which are especially reliable for stratigraphic ordering, show the same succession in the reference section. The biggest differences are found in the species *Acutimitoceras convexum*, which can only be found in the two lowermost horizons of ORBV (ORBV-6 and ORBV-5), but has its FAD in RASC unit 23 and reaches to RASC unit 28. Because the error bar is quite short (1 unit), the result can be assumed to be reliable. When only ORBV is analysed with RASC, the position of *Acutimitoceras convexum* corresponds to the reference section. *Gattendorfia costata* reaches from ORBV-5 to ORBV-2 in the reference section and starts after *Acutimitoceras convexum* (ORBV-6 to ORBV-5). In the RASC analysis *Gattendorfia costata* has its FAD in unit 18 and its LAD in unit 81. It starts earlier than *Acutimitoceras convexum* (FAD in unit 23), but the range is strongly elongated and hence the positioning is less reliable.

Stratigraphical arrangement with only FADs

When only FADs are used for the analysis, the result is a straight line, on which the 52 taxa are positioned (Fig. 32). The longest error bar it to be found with *Nicimitoceras caesari*, except for *Nicimitoceras caesari*, the errors bars do not exceed the length of six units. The two analyses show only minor deviations from each other, except for *Paprothites ruzhencevi*. The taxa, which are reliable for stratigraphical ordering in the analysis with FADs and LADs, show a similar arrangement, but *Paragattendorfia patens* occurs before *Kazakhstania evoluta* and *Eocanites supradevonicus*. In the reference section, the three taxa share the same FAD. The reference section shows a good congruency with this analysis. The major difference is again, that the FAD of *Acutimitoceras convexum* in unit 17 is after the FAD of *Gattendorfia costata* in unit 13.

Fig. 32: Result of the RASC analysis including only FADs with error bars of the Carboniferous dataset.

5.4　Comparison of the results of the three methods

For the comparison of the results of the Unitary Associations, Constrained Optimization and Ranking Scaling analyses, datasets need to be chosen (Tab. 28; Tab. 29):

(1) Unitary Associations: The results of Analysis D (complete corrected datasets) are used for the comparison of the results of the three different methods. In order to achieve a better comparability with the ranked results of the CONOP and RASC analyses, the grouped results of the UA analysis need to be modified. Therefore the "numerical ranges" function in PAST can provide the first and the last UA, in which a species occurs, respectively its FAD and LAD, after which the species can be ranked. In what follows, only the FADs are used.

(2) Constrained Optimization: The CONOP analysis needs to be carried out with FADs and LADs, but for better comparability, only the FADs are used during the further

procedure.

(3) Ranking and Scaling: Only the FADs of the RASC analyses with FADs and LADs are used.

Species	Unitary Associations	Constrained Optimization	Ranking and Scaling	Contradictions
Kosmoclymenia lamellosa *Progonioclymenia acuticostata* *Clymenia laevigata* *Cymaclymenia cordata* *Mimimitoceras pompeckji* *Gonioclymenia speciosa*	UAs 1 and 2	Units 1 to 5	Units 1 to 11	
Ornatoclymenia ornata *Sellaclymenia torleyi*	UA 3	Units 6 to 9	Units 12 to 16	A clear separation between UA 2 and 3 cannot be reproduced by the two other methods.
Cymaclymenia curvicosta *Cymaclymenia tricarinata*	UA 4	Unit 10	Units 19 to 22	
Kosmoclymenia effenbergensis *Piriclymenia piriformis* *Muessenbiaergia diversa* *Muessenbiaergia coronata*	UAs 5 and 6	Units 11 to 17	Units 24 to 30	
Rodachia dorsocostata *Cyrtoclymenia plicata* *Muessenbiargia xenostriata* *Cyrtoclymenia angustiseptata* *Cymaclymenia warsteinensis* *Mimimitoceras liratum* *Cymaclymenia striata* *Linguaclymenia similis*	UA 7	Units 18 to 23	Units 32 to 45	The UAs 7 to 10 are difficult to reproduce by the other two methods: There are many intermixtures.
Cymaclymenia camerata *Kosmoclymenia undulata* *Mimimitoceras geminum*	UA 8	Units 23 to 26	Units 46 to 49	
Muessenbiaergia bisulcata *Kalloclymenia subarmata*	UA 9	Units 27 to 30	Units 50 to 55	
Muessenbiaergia ademmeri *Mimimitoceras fuerstenbergi* *Mimimitoceras trizonatum*	UA 10	Units 31 to 35	Units 56 to 63	
Soliclymenia paradoxa *Glatziella glaucopis* *Effenbergia falx* *Kalloclymenia pessoides* *Parawocklumeria patens* *Parawocklumeria distorta*	UAs 11 and 12	Units 38 to 42	Units 66 to 77	
				...

Species	Unitary Associations	Constrained Optimization	Ranking and Scaling	Contradictions
...				
Kamptoclymenia endogona Kamptoclymenia trigona Effenbergia minutula Mimimitoceras nageli Balvia globulare Lissoclymenia wocklumeri Mimimitoceras lentum Discoclymenia cucullata	UAs 13 to 16	Units 44 to 64	Units 78 to 106	The clear separation of UA 13 is not supported by the other two methods
Parawocklumeria paradoxa Postglatziella carinata Finiclymenia wocklumensis Kenseyoceras nucleus	UAs 17 to 19	Units 65 to 70	Units 108 to 113	
Wocklumeria denckmanni Cymaclymenia involvens Mimimitoceras rotersi	UAs 20 to 22	Units 71 to 78	Units 115 to 119	

Tab. 28: Comparison of the results of the three different methods for the Devonian dataset.

Species	Unitary Associations	Constrained Optimization	Ranking and Scaling	Contradictions
Acutimitoceras prorsum Postclymeina evoluta Acutimitoceras procedens Acutimitoceras stockumense Nicimitoceras carinatum Acutimitoceras kleinerae Acutimitoceras intermedium Acutimitoceras subbilobatum	UA 1	Units 1 to 4	Units 1 to 12	In the RASC analysis, Hasselbachia sphaeroidalis, which has its FAD in UA 3, is grouped with these species.
Nicimitoceras caesari	UA 2	Unit 19	Unit 55	Nicimitoceras caesari forms UA 2. This species is ranked very differently in the other analyses: In CONOP and RASC, Nicimitoceras caesari is grouped with species, which have their FAD in UA 5. This is not plausible, because Nicimitoceras caesari only occurs in the one horizon section SK, which is situated in the lower part of the complete succession.
				...

Species	Unitary Associations	Constrained Optimization	Ranking and Scaling	Contradictions
...				
Acutimitoceras convexum *Acutimitoceras undulatum* *Voehringerites peracutus* *Weyerella reticulum* *Acutimitoceras acutum* *Gattendorfia subinvoluta* *Hasselbachia sphaeroidalis* *Weyerella concava* *Acutimitoceras antecedens* *Globimitoceras globiforme* *Mimimitoceras hoennense* *Nicimitoceras subacre* *Gattendorfia costata* *Nicimitoceras heterolobatum* *Nicimitoceras trochiforme* *Eocanites nodosus* *Gattendorfia tenuis* *Mimimitoceras varicosum*	UA 3	Units 5 to 15	Units 14 to 39	In the RASC analysis, *Acutimitoceras subbilobatum*, which has its FAD in UA 1 is grouped with taxa, which have their FAD in UA 3.
Hasselbachia multisulcata *Acutimitoceras exile* *Paprothites dorsoplanus*	UA 4	Units 14 to 17	Units 33 to 41	The index species *Paprothites dorsoplanus* is grouped with species, which have their FAD in UA 3 in CONOP and RASC.
Paprothites raricostatus *Hasselbachia gracilis* *Costimitoceras ornatum* *Paragattendorfia globiformis* *Weyerella molaris*	UA 5	Units 18 to 19	Units 42 to 57	
Paprothites ruzhencevi	UA 6	Unit 14	Unit 80	Only specimen: *Paprothites ruzhencevi*
Pseudarietites westfalicus *Eocanites brevis* *Pseudarietites subtilis* *Acutimitoceras depressum* *Eocanites spiratissimus* *Eocanites tener*	UAs 7 to 9	Units 21 to 22	Units 58 to 69	
Eocanites carinatus *Nicimitoceras acre* *Acutimitoceras simile* *Gattendorfia crassa*	UA 10	Units 23 to 25	Units 70 to 75	*Acutimitoceras depressum*, which is grouped with these species in the CONOP analysis, has its FAD in UA 8.
Pseudarietites planissimus *Eocanites supradevonicus* *Kazakhstania evoluta* *Paragattendorfia patens*	UAs 11 and 12	Units 29 to 29	Units 78 to 92	
Eocanites planus *Paralytoceras serratum*	UA 13	Units 30 to 31	Units 99 to 100	

Tab. 29: Comparison of the results of the three different methods for the Carboniferous dataset.

5.5 Comparison with the existing ammonoid zonation

Devonian

The existing modern ammonoid zonation, which was used in this study, (Korn 2002) distinguishes nine zones for the late Famennian. These nine zones only partially agree with the stratigraphical succession obtained by the Unitary Associations, Constrained Optimization and Ranking and Scaling methods. Only the results of the RASC method fully coincide with the existing modern ammonoid zones. In the results of the UA method, *Effenbergia lens* has its FAD before *Muessenbiaergia parundulata*. This is neither confirmed by the existing zonation, nor by the fossil content, as *Effenbergia lens* has its FAD after *Muessenbiaergia parundulata* in DASS, E77 and M1. Assumably the addition of ORSTA and ORSTB produces the error, because these two sections only begin within the *Effenbergia lens* Zone. Remarkably, the same data matrix was used to generate the UA and the RASC results. A possible explanation for the earlier FAD of *Effenbergia lens* could be the coexistence of *Effenbergia lens* with very long-ranging species. In the CONOP analysis, *Effenbergia lens* occurs before *Muessenbiaergia parundulata* and *Muessenbiaergia sublaevis*, which underlines that *Effenbergia lens* is very hard to rank with the UA and CONOP methods.

Carboniferous

The existing modern ammonoid zonation (Korn 2002) distinguishes five zones for the early Tournaisian. These five zones perfectly agree with the stratigraphical succession obtained by the Unitary Associations, Constrained Optimization and Ranking and Scaling methods.

5.6 Comparison of the three methods

As argued by Hammer and Harper (2006), which method is best for biostratigraphy depends on the data available and the purpose of the investigation. In my opinion, the format of the available input data is not that important, because with PAST it is very easy to transform samples to events (UA to RASC) and events to samples (RASC to UA) once they are in the right format. Nonetheless, event data is better suited to sections with more than three horizons, because the first and the last horizon do not provide reliable information about a FAD or a LAD. Because the manual data input for CONOP takes a long time and is prone to mistakes, I recommend the R package CONOP9companion by Renaudie (2013), which uses the UA input matrices. On the other hand, the amount of data is very important, because PAST carries out the UA and RASC analyses very quickly and hence can process large amounts of data quickly, on the other hand CONOP needs more time for its analysis and is therefore better suited for smaller datasets.

Hammer and Harper (2006) suggest that RASC is the best method for wells, because in this case only LADs are available for the analysis. Range contradictions and co-occurrence breaking do not pose problems to the RASC method. I also think this method is suitable for other data, because it is very fast and provides reliable results. It is the only method where the results perfectly coincide with the existing ammonoid zonation. Unfortunately a correlation of the different sections is not possible using the Ranking and Scaling method. For event data with FADs and LADs available, Hammer and Harper (2006) recommend CONOP, because it is based on the minimization of range extensions, while honouring co-occurrences and other constraints. However CONOP is labour intensive and does not provide significant advantages over the other two methods, the result is not more reliable and does not show a better stratigraphical succession or resolution than UA and RASC. However, the result mirrors global maximal ranges better and is hence better suitable for comparisons. In this study, this is not necessary, because the area of the sampling localities is limited. For taxa-in-samples data Hammer and Harper (2006) recommend the UA method, although they think CONOP is suitable as well. They consider the UA method, which focuses on observed co-occurrences, as fast and transparent. I agree with their comment and additionally want to stress the fact that the associations facilitate the separation of zones, because changes of the fossil succession are easily recognizable, so that associated UAs can be easily grouped (Tab. 30) .

UA	CONOP	RASC
Carboniferous: 13 units Devonian: 22 units	Carboniferous: 32 units Devonian: 87 units	Carboniferous: 104 units Devonian: 130 units
Resolution low, conservative, considers robustness more important; High lateral reproducibility	Much higher resolution potential than empirical zonations (Cody et al. 2008)	High resolution, but at the cost of losing global originations and extinctions
Carboniferous: agrees with existing zonation Devonian: disagrees with existing zonation	Carboniferous: agrees with existing zonation Devonian: disagrees with existing zonation	Carboniferous: agrees with existing zonation Devonian: agrees with existing zonation
Sequence Carboniferous: *Nicimitoceras caesari* right *Paprothites ruzhencevi*? Partly corresponds to CONOP Sequence Devonian: Partly corresponds to CONOP and RASC in *laevigata-piriformis* Zone and in *endogona-denckmanni* Zone Disagrees with CONOP and RASC in *sublaevis-lens* Zone	Sequence Carboniferous: *Nicimitoceras caesari* wrong *Paprothites ruzhencevi*? Partly corresponds to UA Sequence Devonian: Partly corresponds to UA and RASC in *laevigata-piriformis* Zone and in *endogona-denckmanni* Zone Partly corresponds to RASC in *sublaevis-lens* Zone	Sequence Carboniferous: *Nicimitoceras caesari* wrong *Paprothites ruzhencevi*? Disagrees with CONOP and UA Sequence Devonian: Partly corresponds to UA and CONOP in *laevigata-piriformis* Zone and in *endogona-denckmanni* Zone Partly corresponds to CONOP *sublaevis-lens* Zone

Tab. 30: Differences and similarities of the UA, CONOP and RASC methods.

Comparison of UA and CONOP

Galster et al. (2010) compared the UA and CONOP methods for the stratigraphic succession of Neogene diatoms. They found "... that the UAM is an extremely powerful and unique theory allowing an in-depth analysis of the internal conflicting inter-taxon stratigraphical relationships, inherent to any complex biostratigraphical database". In their opinion the main disadvantage of CONOP is the simulated annealing algorithm, which does not solve cycles in the FADs and LADs. These inter-event cycles have to be calculated separately. On the other hand the UA method treats all events belonging to a cycle as equivalent. They found the implementation of UA in PAST useful for producing a complete analysis of a complex dataset in a short time. The main advantages of the UA method are (1) the detailed analysis of contradictions, (2) the individual comparison of the horizons, (3) the range chart, which displays chronological discontinuities, (4) tools, which allow for the analysis of the diachronism of datums (5) the visualization via graphs and (6) the display of co-occurrences.

Comparison of UA and RASC

Baumgartner (1984) compared the methods UA and RASC for the stratigraphic succession of Mesozoic radiolarians. He had to face the problem, that the radiolarian record is mainly dissolution controlled and therefore has its maximum at the end of its range and hence is non-random. Baumgartner (1984) found that the UA method produces "... maximum ranges of the species relative to each other by stacking co-occurrence data from all sections and therefore compensate for the local dissolution effects". RASC is based on the assumption of a random distribution for this reason it produces shorter average ranges than the UA maximum ranges. Nevertheless some taxa show similar ranges in both results. These species possess a low diversity and are dissolution resistant so they have a consistent occurrence throughout their range and are thus considered more reliable. With my data, scaling has not produced a useful result.

Comparison of CONOP and RASC

Cooper et al. (2001) compared the methods CONOP and RASC for the stratigraphic succession of Paleocene to lower Miocene foraminifera, nannofossils, dinoflagellates, and miospores. They found that ".. the RASC probable sequence and CONOP composite sequence are remarkably similar, and both compare well with classical graphic correlation". Both techniques greatly improve the precision compared to conventional biostratigraphy and show that a higher quality result can be obtained from the same data. The RASC method gives the best estimate of events to be encountered in a new well, the CONOP method, which is based on maximum ranges, is most readily related to existing zonal schemes. In other words, RASC gives the most probable order of events whereas CONOP gives the best approximation of the true stratigraphic range. Additionally, CONOP provides greater precision by an order-of-magnitude. In conclusion, they do not consider the CONOP and the RASC methods as alternatives, but complementary to one another.

6 Discussion

6.1 Suggestion of new biozones

Devonian

The existing zones can be mostly understood in the results of all three analyses (Tab. 31). The existing Devonian ammonoid zonation can be tested by using and comparing different modern biostratigraphical approaches. None of the existing zones can be further subdivided. All separated zones can be easily correlated with the zonation from Korn (2002).

Unitary Association	Biozone	Remarks
UAs 1 and 2	*Clymenia laevigata* Zone	
UAs 3 and 4	*Ornatoclymenia ornata* Zone	Within all three methods, a clear separation of UA 3 and 4 can be proven. The taxa, which occur in UA 4 (*Cymaclymenia curvicosta* and *Cymaclymenia tricarinata*) only posses one occurrence in the E77 section, so that the separation of a new zone is not sensible.
UAs 5 and 6	*Piriclymenia piriformis* Zone	
It is not possible to distinguish the *Muessenbiaergia sublaevis*, *Muessenbiaergia parundulata* and *Effenbergia lens* Zone with the UA method, because they do not show the same succession as in the existing ammonoid zonation. The only method, which mirrored the existing succession, is RASC. Therefore these zones are discriminated by means of the RASC method.		
UAs 7 and 8	*Muessenbiaergia sublaevis* Zone	Units 32 to 49 of RASC. The assignation of the *Muessenbiaergia sublaevis* and the *Muessenbiaergia parundulata* Zone correspond to the results of the UA method, only the *Effenbergia lens* Zone, which is located in the UAs 10 to 12, has a too late occurrence in UA 8.
UA 9	*Muessenbiaergia parundulata* Zone	Units 50 to 55 of RASC
UAs 10 to 12	*Effenbergia lens* Zone	Units 56 to 77 of RASC
UAs 13 to 15	*Kamptoclymenia endogona* Zone	
UAs 16 to 19	*Parawocklumeria paradoxa* Zone	
UAs 20 to 22	*Wocklumeria denckmani* Zone	

Tab 31: Affirmation of the existing modern ammonoid zonation on the basis of the results of the Unitary Associations method.

Carboniferous

The existing zones can be mostly understood in the results of all three analyses (Tab. 32). The existing Carboniferous ammonoid zonation can be tested by using and comparing different modern biostratigraphical approaches. One of the existing zones can be further subdivided. All separated zones can be easily correlated with the zonation from Korn (2002).

Unitary Association	Biozone	Remarks
UAs 1 and 2	*Acutimitoceras prorsum* Zone	
UA 3	*Gattendorfia subinvoluta* Zone	
UA 4	*Paprothites dorsoplanus* Zone	A clear separation of UA 4 and 5 can be seen in the results of all three analyses, so that it is possible to distinguish two zones: The *Paprothites dorsoplanus* Zone equals UA 4.
UAs 5 to 6	*Weyerella molaris* Zone	The new *Weyerella molaris* Zone equals UA 5. *Weyerella molaris* is chosen as an index fossil, because it possesses the most occurrences (13) in the most sections (4) of all species in this zone. UA 6 only includes *Paprothites ruzhencevi*.
UAs 7 to 10	*Pseudarietites westfalicus* Zone	A clear subdivision can be seen between UA 9 and 10 in the results of all three analyses. Nonetheless, the species, which form UA 10, only occur in the Oberrödinghausen locality and hence are not suitable as index fossils for a new zone
UAs 11 to 13	*Paragattendorfia patens* Zone	A clear separation between UA 12 and 13 can be seen. However it is not sensible to subdivide the *Paragattendorfia patens* Zone, because the two taxa in UA 13 (*Eocanites planus* and *Paralytoceras serratum*) have their only occurrence in the horizon ORBV-1.

Tab. 32: Suggestion of new biozones on the basis of the results of the Unitary Assocations method.

7 Summary

This MSc thesis not only seeks to refine the latest Devonian and earliest Carboniferous ammonoid stratigraphy for the Rhenish Mountains, but also to clarify if the modern biostratigraphical methods Unitary Associations (UA), Constrained Optimization (CONOP) and Ranking and Scaling (RASC) are qualified for this purpose as well as which method is best suitable.

Although the UA method was carried out with seven different datasets, the complete corrected dataset proved to be the most reliable. Neither the use of first occurrences only nor the omission of singletons improves the result. All of the ammonoid associations found in the Devonian and Carboniferous analyses matched the associations in the reference section. The result of the CONOP analysis coincides largely with the reference section for the Devonian and almost completely with the reference section for the Carboniferous. The results of the RASC analysis show a perfect fit for the Devonian as well as for the Carboniferous reference sections. The analysis with FADs and LADs shows only minor deviations from the analysis with only FADs.

Principally, the UA, CONOP and RASC methods lead to similar results with respect to the succession of occurrence events of the analysed ammonoid species in the various sections. It is possible to understand the same succession and grouping of species in the results of the UA, the CONOP and the RASC method. Of the three approaches, the UA method results in the lowest resolution, but it has the highest robustness. Application of the CONOP method results in a higher but less robust resolution and the seemingly highest resolution is provided by RASC. For the Devonian, only the results of the RASC method coincides with the existing modern ammonoid zonation, the *Effenbergia lens* Zone, the *Muessenbiaergia parundulata* Zone and the *Muessenbiaergia sublaevis* Zone can not be resolved by the UA and the CONOP methods. For the Carboniferous the results of all methods coincide with the existing modern ammonoid zonation.

On the basis of the results of the three analyses new biozones for the latest Famennian and earliest Tournaisian are suggested. The existing modern ammonoid zonation for the Devonian was confirmed by all three methods, for the Carboniferous, a slight refinement is needed. In addition to the existing modern ammonoid zonation the separation of UA 4 and UA 5 was revealed in the results of all three analyses. UA 4 forms the existing *Paprothites dorsoplanus* Zone, UA 5 forms the new *Weyerella molaris* Zone.

As already stated by former studies, which method is most suitable depends on the data available and the purpose of the investigation. Out of the three biostratigraphical analysis methods I consider the CONOP approach to be the least suitable, because the data input and the calculation itself takes a long time. UA and RASC are recommendable, because they are fast and yield good results. The conservative UA method groups taxa, which yields a lower resolution but facilitates the separation of zones. The RASC method on the other hand shows a seemingly high resolution. It is also the only method, whose result perfectly mirrors the existing modern ammonoid zonation used in this thesis. Unfortunately the correlation of the different profiles is not possible using the RASC method, so the two approaches Unitary Associations and Ranking and Scaling should be used together.

So far, applications of the Unitary Associations method have been published for Middle Devonian, Early Triassic, Middle Triassic and Late Cretaceous ammonoid associations. The possibilities of the application of the Ranking and Scaling method to ammonoid biostratigraphy have to date not been published. There are many time intervals left where the possibilities of a refinement of the existing ammonoid zonation with the Unitary Associations method, the Ranking and Scaling method or a combination of them can be tested. Furthermore, it can be tested, whether the combination of the Unitary Associations method and the Ranking and Scaling method leads to a refinement of the biostratigraphy of other fossil groups.

8 References

Agterberg, F. P., and F. M. Gradstein. 1999. "The RASC Method for Ranking and Scaling of Biostratigraphic Events." *Earth-Science Reviews* 46: 1–25.

Alberti, H., H. Groos-Uffenorde, M. Streel, H. Uffenorde, and O. H. Walliser. 1974. "The Stratigraphical Significance of the Protognathodus Fauna from Stockum (Devonian/Carboniferous Boundary, Rhenish Schiefergebirge)." *Newsletters on Stratigraphy* 3: 263–76.

Bartzsch, K., and D. Weyer. 1987. "Die Unterkarbonische Ammonoidea-Tribus Pseudarietitini." *Abhandlungen und Berichte für Naturkunde und Vorgeschichte* 13: 59–69.

Baumgartner, P. O. 1984. "Comparison of Unitary Associations and Probabilistic Ranking and Scaling as applied to Mesozoic Radiolarians." *Computers & Geosciences* 10 (1): 167–83.

Becker, R. T. 1988. "Ammonoids from the Devonian-Carboniferous Boundary in the Hasselbach Valley (Northern Rhenish Slate Mountains)." *Courier Forschungsinstitut Senckenberg* 1: 193–213.

———. 1993. "Anoxia, eustatic Changes, and Upper Devonian to Lowermost Carboniferous global Ammonoid Diversity." *In: House, M.R. (ed.): The Ammonoidea: Environment, Ecology, and Evolutionary Change, Systematics Association Special Volume* 47: 115–63.

———. 1996. "New Faunal Records and Holostratigraphic Correlation of the Hasselbachtal D/C-Boundary Auxiliary Stratotype (Germany)." *Annales de La Société Géologique de Belgique* 117: 19–45.

Becker, R. T., M. J. M. Bless, C. Brauckmann, L. Friman, K. Higgs, H. Keupp, D. Korn, et al. 1984. "Hasselbachtal, the Section best displaying the Devonian-Carboniferous Boundary Beds in the Rhenish Massif (Rheinisches Schiefergebirge)." *Courier Forschungsinstitut Senckenberg* 67: 181–91.

Bockwinkel, J., and V. Ebbighausen. 2006. "A new Ammonoid Fauna from the *Gattendorfia-Eocanites* Genozone of the Anti-Atlas (Early Carboniferous; Morocco)." *Fossil Record* 9: 87–129.

Brühwiler, T., H. Bucher, A. Brayard, and N. Goudemand. 2010. "High-Resolution Biochronology and Diversity Dynamics of the Early Triassic Ammonoid Recovery: The Smithian Faunas of the Northern Indian Margin." *Palaeogeography, Palaeoclimatology, Palaeoecology* 297 (2): 491–501.

Claoué-Long, J. C. 1995. "Two Carboniferous Ages: A Comparison of SHRIMP Zircon Dating with Conventional Zircon Ages and 40Ar/39Ar Analysis." *SEPM Special Publication* 54: 1–21.

Claoué-Long, J. C., P. J. Jones, J. Roberts, and S. Maxwell. 1992. "The numerical Age of the Devonian-Carboniferous Boundary." *Geological Magazine* 129 (3): 281–91.

Clausen, C.-D., D. Korn, R. Feist, K. Leuschner, H. Groos-Uffenorde, F. W. Luppold, D. Stoppel, K. Higgs, and M. Streel. 1994. "Die Devon/Karbon-Grenze bei Stockum (Rheinisches Schiefergebirge)." Geologie und Paläontologie in Westfalen 29: 71–95.

Clausen, C.-D., and D. Korn. 2008. "Höheres Mitteldevon und Oberdevon des nördlichen Rheinischen Schiefergebirges (mit Velberter Sattel und Kellerwald)." 52: 439–81.

Cody, R. D., R. H. Mapes, David M. H., and P. M. Sadler. 2008. "Thinking Outside the Zone: High-Resolution Quantitative Diatom Biochronology for the Antarctic Neogene." Palaeogeography, Palaeoclimatology, Palaeoecology 260 (1): 92–121.

Cooper, R. A., J. S. Crampton, J. I. Raine, F. M. Gradstein, H. E. G. Morgans, P. M. Sadler, C. P. Strong, D. Waghorn, and G. J. Wilson. 2001. "Quantitative Biostratigraphy of the Taranaki Basin, New Zealand: A Deterministic and Probabilistic Approach." *AAPG Bulletin* 85 (8): 1469–98.

Denckmann, A. 1894. "Zur Stratigraphie des Oberdevon im Kellerwalde und einigen benachbarten Devon-Gebieten." *Jahrbuch der Preußischen Geologischen Landesanstalt* 15: 8–64.

———. 1901. "Ueber das Oberdevon auf Blatt Balve (Sauerland)." *Jahrbuch der Königlich Preussischen Geologischen Landesanstalt und Bergakademie* 21: I–XIX.

Frech, F. 1897. "Lethaea Geognostica Oder Beschreibung und Abbildung der für die Gebirgs-Formationen bezeichnendsten Versteinerungen. I. Theil. Lethaea Palaeozoica. 2. Band."

———. 1902. "Über Devonische Ammoneen." *Beiträge Zur Paläontologie Österreich-Ungarns und des Orients* 14: 27–112.

Galster, F., J. Guex, and Ø. Hammer. 2010. "Neogene Biochronology of Antarctic Diatoms: A Comparison between two quantitative Approaches, CONOP and UAgraph." *Palaeogeography, Palaeoclimatology, Palaeoecology* 285 (3): 237–47.

Gradstein, F. M., and F. P. Agterberg. 1982. "Models of Cenozoic Foraminiferal Stratigraphy—northwestern Atlantic Margin." In *Quantitative Stratigraphic Correlation.*, edited by J. M. Cubitt and R. A. Reyment, 119–73. Chichester: Wiley.

Gradstein, F. M., F. P. Agterberg, J. C. Brower, and W. Schwarzacher. 1985. *Quantitative Stratigraphy.* Paris: Reidel, Dordrecht and UNESCO.

Gradstein, F. M., J. G. Ogg, M. D. Schmitz, and G. M. Ogg. 2012. *The Geologic Time Scale 2012.* 1st ed. Elsevier.

Guex, J. 1991. *Biochronological Correlations.* Berlin: Springer Verlag.

Hammer, Ø. 2012. "PAST PAleontological STatistics Version 2.17 Reference Manual."

Hammer, Ø., and D. A. T. Harper. 2001. "PAST: Paleontological Statistics Software Package for Education and Data Analysis." *Palaeontologica Electronica* 4 (1): 1–9.

———. 2006. *Paleontological Data Analysis.* Blackwell Publishing.

Hay, W. W. 1972. "Probabilistic Stratigraphy." *Eclogae Geologicae Helvetiae* 65: 255–66.

Henke, W. 1924. "Erläuterungen zu Blatt Endorf." *Geologische Karte von Preußen und Benachbarten Bundesstaaten 1:25 000* Lieferung 236: 1–44.

House, M. R. 2002. "Strength, Timing, Setting and Cause of Mid-Palaeozoic Extinctions." *Palaeogeography, Palaeoclimatology, Palaeoecology* 181: 5–25.

Kayser, E. 1872. "Studien aus dem Rheinischen Devon. III. Die Fauna des Rotheisensteins von Brilon in Westfalen." *Zeitschrift der Deutschen Geologischen Gesellschaft* 24: 653–90.

———. 1873. "Ueber die Fauna des Nierenkalkes vom Enkeberge und der Schiefer von Nehden bei Brilon, und ueber die Gliederung des Oberdevon im Rheinischen Schiefergebirge." *Zeitschrift der Deutschen Geologischen Gesellschaft* 25: 602–74.

Kemple, W. G., P. M. Sadler, and D. J. Strauss. 1989. "A Prototype Constrained Optimization Solution to the Time Correlation Problem." In *Statistical Applications in the Earth Sciences*, edited by F. P. Agterberg and G. F. Bonham-Carter, 9: 417–25. Geological Survey of Canada Paper 89.

———. 1995. "Extending Graphic Correlation to Many Dimensions: Stratigraphic Correlation as Constrained Optimization." In *Extending Graphic Correlation to Many Dimensions: Stratigraphic Correlation as Constrained Optimization*, edited by K. O. Mann and H. R. Lane, 53: 65–82. SEPM Sp. Pap.

Klein, C., & Korn, D. (2016). Quantitative analysis of the late Famennian and early Tournaisian ammonoid stratigraphy. *Newsletters on   Stratigraphy, 49*(1), 1-26.

Korn, D. 1981. "Ein neues, Ammonoiden-führendes Profil an der Devon-Karbon-Grenze im Sauerland (Rhein. Schiefergebirge)." *Neues Jahrbuch Für Geologie Und Paläontologie, Monatshefte* 1981(9): 513–26.

———. 1984. "Die Goniatiten der Stockumer *Imitoceras*-Kalklinsen (Ammonoidea; Devon/Karbon-Grenze)." *Courier Forschungsinstitut Senckenberg* 67: 71–89.

———. 1986. "Ammonoid Evolution in Late Famennian and Early Tournaisian." *Annales de La Société Géologique de Belgique* 109 (Fascicule 1 (Late Devonian events around the Old Red Continent)).

———. 1991. "Threedimensionally preserved Clymeniids from the Hangenberg Black Shale of Drewer (Cephalopoda, Ammonoidea; Devonian-Carboniferous Boundary; Rhenish Massif)." *Neues Jahrbuch für Geologie und Paläontologie, Monatshefte* 1991(9): 553–63.

———. 1993. "The Ammonoid Faunal Change near the Devonian-Carboniferous Boundary." *Annales de La Société Géologique de Belgique* 115: 581–93.

———. 1994. "Devonische und Karbonische Prionoceraten (Cephalopoda, Ammonoidea) aus dem Rheinischen Schiefergebirge." *Geologie und Paläontologie in Westfalen* 30: 1–85.

———. 2002. "Historical Subdivisions of the Middle and Late Devonian Sedimentary Rocks in the Rhenish Mountains by Ammonoid Faunas." *Senckenbergiana Lethaea* 82: 545–55.

Korn, D., Z. Belka, S. Fröhlich, M. Rücklin, and J. Wendt. 2004. "The youngest African Clymeniids (Ammonoidea, Late Devonian) - Failed Survivors of the Hangenberg Event." *Lethaia* 37: 307–15.

Korn, D., C. D. Clausen, Z. Belka, K. Leuteritz, F. W. Luppold, R. Feist, and D. Weyer. 1994. "Die Devon/Karbon- Grenze bei Drewer (Rheinisches Schiefergebirge)." *Geologie und Palaontologie in Westfalen* 34: 97–147.

Korn, D., and A. Ilg. 2007. "AMMON. Database of Palaeozoic Ammonoidea." http://www.wahre-staerke.com/ammon/.

Korn, D., and C. Klug. 2002. *Ammoneae Devonicae.* Vol. 138. Fossilium Catalogus, I: Animalia. Backhuys.

————. 2012. "Palaeozoic Ammonoids – Diversity and Development of Conch Morphology." In *Extinction Intervals and Biogeographic Perturbations through Time*, edited by John Talent, 491–534. Berlin: Springer.

Korn, D., and F. W. Luppold. 1987. "Nach Clymenien und Conodonten gegliederte Profile des Oberen Famennium Im Rheinischen Schiefergebirge." *Courier Forschungsinstitut Senckenberg* 92: 199–223.

Korn, D., and D. Weyer. 2003. "High Resolution Stratigraphy of the Devonian-Carboniferous Transitional Beds in the Rhenish Mountains." *Mitteilungen aus dem Museum für Naturkunde in Berlin, Geowissenschaftliche Reihe* 6: 79–124.

Lange, W. 1929. "Zur Kenntnis des Oberdevons am Enkeberg und bei Balve (Sauerland)." *Abhandlungen Der Preußischen Geologischen Landesanstalt, Neue Folge* 119: 1–132.

Luppold, F. W., C.-D. Clausen, D. Korn, and D. Stoppel. 1994. "Devon/Karbon-Grenzprofile im Bereich von Remscheid-Altenaer Sattel, Warsteiner Sattel, Briloner Sattel und Attendorn-Elsper Doppelmulde (Rheinisches Schiefergebirge)." *Geologie und Paläontologie in Westfalen* 29: 7–69.

Luppold, F. W., G. Hahn, and D. Korn. 1984. "Trilobiten-, Ammonoideen- und Conodonten-Stratigraphie des Devon/Karbon-Grenzprofiles auf dem Müssenberg (Rheinisches Schiefergebirge)." *Courier Forschungsinstitut Senckenberg* 67: 91–121.

Monnet, C., and H. Bucher. 1999. "Biochronologie Quantitative (associations Unitaires) des Faunes d'Ammonites du Cenomanien du Sud-Est de la France." *Bulletin de La Société Géologique de France* 170: 599–610.

————. 2002. "Cenomanian (early Late Cretaceous) Ammonoid Faunas of Western Europe. Part I: Biochronology (unitary Associations) and Diachronism of Datums." *Eclogae Geologicae Helvetiae* 95 (1): 57–74.

————. 2006. "Anisian (Middle Triassic) Ammonoids from North America: Quantitative Biochronology and Biodiversity." *Stratigraphy* 2 (4): 311–26.

Monnet, C., C. Klug, N. Goudemand, K. De Baets, and H. Bucher. 2011. "Quantitative Biochronology of Devonian Ammonoids from Morocco and Proposals for a refined Unitary Association Method." *Lethaia* 44 (4): 469–89.

Paproth, E., and M. Streel. 1982. "Devonian–Carboniferous Transitional Beds of the Northern 'Rheinisches Schiefergebirge.'" *IUGS Working Group on the Dev./Carb. Boundary.*

Price, J. D., and M. R. House. 1984. "Ammonoids near the Devonian-Carboniferous Boundary." *Courier Forschungsinstitut Senckenberg* 67: 15–22.

Regents of the University of California. 2011. "Stratigraphic Correlation and Seriation." https://ilearn.ucr.edu/webapps/portal/frameset.jsp?tab_group=courses&url=%2Fwebapps%2Fblackboard%2Fexecute%2FcourseMain%3Fcourse_id%3D_136576_1.

Renaudie, J. 2013. "Tools for Integrating Biostratigraphic Software CONOP9 in a Statistical Workflow. (R Package)." https://github.com/plannapus/CONOP9companion.

Sadler, P. M. 2009. "Constrained Optimization Approaches to the Paleobiologic Correlation and Seriation Problems: A User's Guide and Reference Manual to the CONOP Program Family."

Savary, J., and J. Guex. 1999. "Discrete Biochronological Scales and Unitary Associations: Description of the BioGraph Computer Program." *Meomoires de Geologie (Lausanne)* 34.

Schindewolf, O. H. 1937. "Zur Stratigraphie und Paläontologie der Wocklumer Schichten (Oberdevon)." *Abhandlungen der Preußischen Geologischen Landesanstalt, Neue Folge* 178: 1–132.

Schmidt, H. 1922. "Das Oberdevon-Culm-Gebiet von Warstein i.W. und Belecke." *Jahrbuch der Preußischen Geologischen Landesanstalt* 41(1920): 254–339.

————. 1924. "Zwei Cephalopodenfaunen an der Devon-Carbongrenze im Sauerland." *Jahrbuch der Preußischen Geologischen Landesanstalt* 44(1923): 98–171.

Trapp, E., B. Kaufmann, K. Mezger, D. Weyer, and D. Korn. 2004. "Numerical Calibration of the Devonian-Carboniferous Boundary: Two new U-Pb Isotope Dilution-Thermal Ionization Mass Spectrometry Single-Zircon Ages from Hasselbachtal (Sauerland, Germany)." *Geology* 32: 857–60.

Vöhringer, E. 1960. "Die Goniatiten der Unterkarbonischen Gattendorfia-Stufe im Hönnetal (Sauerland)." *Fortschritte in der Geologie von Rheinland und Westfalen* 3: 107–96.

Von Buch, L. 1832. "Über Goniatiten." *Physikalische Abhandlungen der Königlichen Akademie der Wissenschaften Berlin* 1831: 159–87.

Wedekind, R. 1914. "Monographie der Clymenien des Rheinischen Gebirges." *Abhandlungen der Gesellschaft der Wissenschaften in Göttingen, Mathematisch-Physikalische Klasse, Neue Folge* 10(1): 1–73.

Ziegler, W. 1962. *Taxionomie und Phylogenie Oberdevonischer Conodonten und ihre Stratigraphische Bedeutung.* Hessisches Landesamt für Bodenforschung. Abhandlungen 87: 7–77 .

———. 1971. "Post-Symposium Excursion, Sept. 15–18, 1971, to Rhenish Slate Mountains and Hartz Mountains. A Field Trip Guidebook." In *Symposium on Conodont Taxonomy Marburg/Lahn, Sept,* 4:1–47.

Printed in the United States
By Bookmasters